Advances in World Aquaculture, Volume 4
Managing Editor, Paul A. Sandifer

Shrimp Culture in North America and the Caribbean

Edited by

Paul A. Sandifer

Marine Resources Division
South Carolina Wildlife and
Marine Resources Department
P.O. Box 12559
Charleston, SC 29412

Published by

THE WORLD AQUACULTURE SOCIETY

The World Aquaculture Society
143 J.M. Parker Coliseum
Louisiana State University
Baton Rouge, LA 70803

Library of Congress Catalog Number: 91-065427

ISBN: 0-9624529-3-9

PREFACE

Over the last 20 years, aquaculture has become an important means of producing shrimp in many areas of the world, especially in Southeast Asia and Latin America. Today, farms account for more than 25% of the total volume of world shrimp production. However, while much of the technology for shrimp farming has been developed in North America, there has been relatively little commercial development here. Nevertheless, there have been sporadic attempts to develop commercial shrimp farming in various areas of North America for more than 30 years. Recent years have seen a resurgence of interest in commercial shrimp farming in the United States, the Caribbean islands, and Mexico and, although the industry is still quite small, it appears to be on its way to becoming a significant additional source of domestic shrimp production.

Because of the growing interest in shrimp farming in North America, a special World Aquaculture Society session on "Shrimp Culture in North America and the Caribbean" was convened by the Editor at Aquaculture '89 in Los Angeles, California. The purpose of the session was threefold: (1) to review the history and current status of shrimp farming activities in relevant areas of North America and the Caribbean islands; (2) to present some areas of technological advances that appear to hold particular promise for shrimp aquaculture in these geographic areas; and (3) to consider, via examples, the economic environment for shrimp farming. The papers from that session, augmented with a description of the shrimp farming industry in Mexico and a case study of semi-intensification of shrimp farming in Ecuador, comprise the contents of this volume. The final paper is included to provide one view of the development potential of shrimp farming from the private sector. The Editor thanks the authors and the anonymous reviewers who helped to improve the final papers.

Paul A. Sandifer, Charleston, SC. 15 February 1991.

CONTENTS

LIST OF AUTHORS (* denotes senior author)

Alston, Dallas E. Department of Marine Biology, University of Puerto Rico, Mayaguez Campus, P. O. Box 5000, Mayaguez, Puerto Rico

Arthur, Richard H. Aquacultural Concepts, Inc., P. O. Box 40386, San Juan, Puerto Rico 00940-0386

Barck, Lorena E. Hawaii Aquaculture Company, Inc., Kaimuki Technology Enterprise Center, 1103 9th Avenue, Suite 206, Honolulu, Hawaii 96816

*Chamberlain, George. W. Texas Agricultural Extension Service, Route 2, Box 589, Corpus Christi, Texas 78410 (Present address: Director, Aquaculture Research and Technology Management, Ralston Purina International, Checkerboard Square, St. Louis, MO 63164

Cotsapas, Linos. Research Planning, Inc., 1200 Park Street, P.O. Box 328, Columbia, South Carolina 29202

*D'Abramo, Louis R. Department of Wildlife and Fisheries, Mississippi State University, P. O. Drawer, LW, Mississippi State, Mississippi 39762

Daniels, William H. Department of Wildlife and Fisheries, Mississippi State University, P. O. Drawer, LW, Mississippi State, Mississippi 39762

Desmond, Thomas S. Hawaii Aquaculture Company, Inc., Kaimuki Technology Enterprise Center, 1103 9th Avenue, Suite 206, Honolulu, Hawaii 96816

*French, Bruce J. RPI International, Inc., 1200 Park Street, P.O. Box 328 Columbia, South Carolina 29202 (Present address: 2201 Monticello Drive, Tallahassee, Florida 32303)

Fuller, Marty J. Department of Agricultural Economics, Mississippi State University, Box 5187, Mississippi State, Mississippi 39762

Garmendia Nunez, Ernesto A. Direccion General de Acuacultura - Sepesca, Av. Constituyentes Esq. Plan de San Luis Frac., Constitucion, Edif. Salomon Pachuca, Hidalgo 42000, Mexico

Hayes, Miles O. Research Planning, Inc., 1200 Park Street, P. O. Box 328, Columbia, South Carolina 29202

*Hedgecock, Dennis. Aquaculture and Fisheries Program, University of California-Davis, Bodega Marine Laboratory, Bodega Bay, California 94923 and Hawaii Aquaculture Company, Inc., Kaimuki Technology Enterprise Center, 1103 9th Avenue, Suite 206, Honolulu, Hawaii 96816

Heinen, John M. Zeigler Brothers, Inc., Spring and Groundwater Resources Institute, P. O. Box 1746, Shepherdstown, West Virginia 25443

*Hopkins, J. Stephen. James M. Waddell, Jr. Mariculture Research and Development Center, P. O. Box 809, Bluffton, South Carolina 29910

Kohnke, Gil. Pacific Business Center Program, University of Hawaii, Honolulu, Hawaii 96822

MacMichael, Elizabeth R. Hawaii Aquaculture Company, Inc., Kaimuki Technology Enterprise Center, 1103 9th Avenue, Suite 206, Honolulu, Hawaii 96816

*Malecha, Spencer R. Hawaii Aquaculture Company, Inc., Kaimuki Technology Enterprise Center, 1103 9th Avenue, Suite 206, Honolulu, Hawaii 96816

*Pruder, Gary D. The Oceanic Institute, Makapuu Point, P. O. Box 25280, Honolulu, Hawaii 96825

*Rhodes, Raymond J. Division of Marine Resources, South Carolina Wildlife and Marine Resources Department, P. O. Box 12559, Charleston, South Carolina 29412

Roberts, Jonathan. Pacific Business Center Program. University of Hawaii, Honolulu, Hawaii 96822

*Sandifer, Paul A. Division of Marine Resources, South Carolina Wildlife and Marine Resources Department, P. O. Box 12559, Charleston, South Carolina 29412 and James M. Waddell, Jr. Mariculture Research and Development Center, P. O. Box 809, Bluffton, South Carolina 29910

*Shleser, Robert A. Aquacultural Concepts, Inc., P. O. Box 40386, San Juan, Puerto Rico 00940-0386

Smiley, Robert A. James M. Waddell, Jr. Mariculture Research and Development Center, P. O. Box 809, Bluffton, South Carolina 29910 (Present address: Edisto Aquatic Farms, P. O. Drawer 589, Orangeburg, South Carolina 29115)

Stokes, Alvin D. James M. Waddell, Jr. Mariculture Research and Development Center, P. O. Box 809, Bluffton, South Carolina 29910

PART I:

STATUS AND POTENTIAL

STATUS AND OPPORTUNITIES OF SHRIMP FARMING IN THE CARIBBEAN ISLANDS

Robert A. Shleser, Dallas E. Alston, and Richard H. Arthur

ABSTRACT

An overview of freshwater prawn and marine shrimp farming in the Caribbean Islands is presented. The review of status by country shows that most activities are small (less than 100 acres). The farms have primarily been established by small entrepreneurial groups rather than larger commercial investors. In many cases, the activities represent the first commercial aquaculture endeavor for the developers. Both semi-intensive and extensive approaches to farming have been undertaken. Lack of indigenous species suitable for farming has placed an emphasis on providing reliable supplies of postlarvae from maturation facilities or by contractual relationships with larger producers or commercial hatcheries.

Total production in the islands of the Caribbean in 1988 was estimated to be 2,635 MT. This represents only 0.2% of the total shrimp* imported to the United States annually.

Some research and extension support exists on all the major islands. A primary problem hindering the growth of the industry appears to be accessing suitable land and completing a lengthy and poorly defined permitting process.

Prospects for the future are uncertain, given the need to compete with Latin America and Southeast Asian producers for export markets.

* In this report the term shrimp refers to marine shrimp and freshwater prawns (*Macrobrachium*).

INTRODUCTION

In this report, we have attempted to summarize the situation regarding shrimp farming in the Caribbean. The summary is presented by region in order to provide coherence. Regions are reviewed as follows:

The Northern Caribbean or Greater Antilles:
Cuba, The Bahamas, Jamaica, Haiti, Dominican Republic and Puerto Rico;
The Leeward Islands:
St. Thomas, St. Croix, St. Kitts, Antigua and Barbuda;
The Windward Islands:
Dominica, St. Lucia, St. Vincent, Grenada, Barbados, Trinidad and Tobago.
The French West Indies:
Guadaloupe and Martinique.

A primary reference used in developing this paper is an excellent report completed in 1988 by the Foreign Fisheries Analysis Branch of the National Marine Fisheries Service entitled Latin American Shrimp Culture Industry 1986-1990. This document provided a summary of activities in the Caribbean.

In contrast to other geographical areas or even countries, shrimp farming in the Caribbean is very limited. There are a very few examples of commercially competitive shrimp farms in the region. The Caribbean might best be characterized as "having great potential" or "in an early stage of development." At present, there are approximately 1,250 ha of *Macrobrachium* farms and about 860 ha of marine shrimp farms in operation in the Caribbean. Also there are many other farms in various stages of development. Tables 1 and 2 summarize the status of marine shrimp and freshwater prawn farms in the Caribbean Islands.

Uncertainty regarding public policy, permits and funding makes it difficult to place a timetable on aquaculture development or estimate probability of success of the projects underway in the area.

Table 1. Status of marine shrimp farms in the Caribbean.

Country	COMMERCIAL FARMS				HATCHERIES		
	No.of Farms	Total Area (Ha)	Kg/Ha	Yield (MT)	Potential Area (Ha)	Number	Yield Per Mth (million)
Antigua	1	1.6	N/A	N/A	N/A	1	0
Bahamas	0	0	0	0	N/A	0	0
Barbados	0	0	0	0	N/A	0	0
Cuba	N/A	700	1,600	2,286	25,000	3	N/A
Dominica	0	0	0	0	N/A	0	0
D.R.	3	154	1,000	227	8,000	2	2
Grenada	0	0	0	0	N/A	0	0
Guadeloupe	0	0	0	0	N/A	0	0
Haiti	0	0	0	0	8,000	0	0
Jamaica	0	0	0	0	1,200	0	0
Martinique	0	0	0	0	N/A	0	0
Puerto Rico	1	0.1	N/A	N/A	1,500	0	0
St. Croix	0	0	0	0	N/A	0	0
St. Kitts	1	3	N/A	1,362	N/A	1	1
St. Lucia	0	0	0	0	N/A	0	0
St. Vincent	0	0	0	0	N/A	0	0
Trinidad	1	1.6	N/A	N/A	20,235	0	0
TOTAL:	7	860	2,535	3,875	63,935	7	3

GREATER ANTILLES

Cuba

Cuba is believed to have the greatest potential for marine shrimp and freshwater prawn farming of any island in the Caribbean. A rough estimate suggests that more than 25,000 ha of land may be suitable for shrimp farming.

Table 2. Status of freshwater prawn farms in the Caribbean

Country	COMMERCIAL FARMS				HATCHERIES		
	No.of Farms	Total Area (Ha)	Kg/Ha	Yield (MT)	Potential Area (Ha)	Number	Yield Per Mth (million)
Antigua	0	0	0	0	N/A	0	0
Bahamas	0	0	0	0	N/A	0	0
Barbados	0	0	0	0	N/A	0	0
Cuba	0	0	0	0	N/A	0	0
Dominica	N/A	N/A	N/A	N/A	N/A	N/A	N/A
D.R.	28	900	280	311	N/A	12	N/A
Grenada	1	0.2	N/A	N/A	N/A	0	0
Guadeloupe	19	60	65	1,500	N/A	1	2.5
Haiti	0	0	0	0	N/A	0	0
Jamaica	2	38	13.6	355	1,200	1	3
Martinique	60	200	50	250	400	1	1
Puerto Rico	1	48	114	2,375	N/A	2	3
St. Croix	0	0	0	0	N/A	0	0
St. Kitts	0	0	0	0	N/A	0	0
St. Lucia	0	0	0	0	N/A	1	0
St. Vincent	1	0.6	N/A	N/A	N/A	0	0
Trinidad	0	0	0	0	N/A	0	0
TOTAL:	112	1,245.8	522.6	4,791	1,600	18	9.5

At present, the total area used for production of marine shrimp is approximately 700 ha and the farms are producing *Penaeus schmitti*. The farms are located in the Provincia Gramma in the south-central part of the country. Annual production is estimated to be about 570 MT in 1988, with farms averaging 0.6 tons/ha (Rosenberry 1989).

Three hatcheries exist in the country at present. The hatcheries have capacities of 10, 30 and 100 million postlarvae/mo., respectively. At this time, postlarvae production is only sufficient to supply the areas in production.

The farms in Cuba appear to be affected by two problems; providing feed of sufficient quality to obtain optimum growth performance and controlling salinity during the dry season.

"One Cuban specialist believes that the National Aquaculture Enterprise (ENACUI) will begin to make an important contribution to the island shrimp industry by 1990. ENACUI has a special unit working on shrimp culture, the Unidad Presupuestada de Camaronicultura. Cuban technicians have been trained in France, Japan, and the Philippines. The Cubans have worked primarily with *P. schmitti*." (NOAA 1988)

Bahamas

At present there are no commercial shrimp farming activities in the Bahamas. The Morton Company experimented with shrimp farming in seawater pumped into salt ponds on Great Inagua Island for almost 8 years (NOAA 1988). This operation was suspended in 1987.

Worldwide Protein, a subsidiary of Maritec Corporation of Corpus Cristi, Texas, purchased 7,800 ha of salt ponds from Diamond Crystal Salt Company on Long Island, Bahamas (NOAA 1988). By September 1987, operations were discontinued, apparently without having produced any measurable amount of shrimp (Weidner and Niemeier 1988).

Jamaica

Marine Shrimp Farming

Serious interest in marine shrimp farming in Jamaica began in 1987. Three companies have developed plans to build marine shrimp farms.

Land managed by Agro 21 Corporation, an agency of the Jamaican government, has been committed to two projects. The sites are located on the south coast in the region of Mitchelltown bordering the Rio Minho. A 121 ha site will to be developed by the Northern Caribbean Shrimp Company of Puerto Rico in a joint venture with a Jamaican partner. This group proposes to produce *Penaeus vannamei* under semi-intensive conditions.

Two additional shrimp activities are proposed to be established in the same region. One, a joint venture with a Scandinavian group, will develop 81 ha of ponds in the area of Old Harbor. A third activity of approximately 41 ha has been proposed in the vicinity of Spanish Town on the south coast.

Freshwater Prawn Farming

Ocean Protein developed one of the first freshwater prawn farms in the region with a pilot facility in 1971. During 1971-72 16 0.4-ha ponds were completed. A hatchery with a capacity of about 3 million postlarvae/mo. was developed by the end of 1972. By 1974, it was determined that the venture would not meet the investors' needs and the effort was terminated. From 1975 to the present, there have been a number of efforts to re-establish production of prawns and tilapia at this site. Apparently, the financial resources were not adequate to sustain a full development.

Aquaculture Jamaica Limited (AJL) was begun in 1983 (NOAA 1988). This company produces *Macrobrachium* and tilapia in 32 ha of ponds. This farm operates its own hatchery with a capacity to produce 1 million postlarvae/mo. The hatchery reportedly achieves survival rates ranging from 60 to 80 % and harvested 35-60 PL's/l.

The farms producing *Macrobrachium* also produce tilapia on the same farm and in some cases in the same ponds. The production of *Macrobrachium* in Jamaica, from 1985-88, was as follows:

1985	19,091 kg
1986	15,500 kg
1987	8,471 kg
1988	13,636 kg

The government of Jamaica has made 1,200 ha of land available for the development of aquaculture (Wint 1988).

Haiti

Marine Shrimp Farming

At present, there are no marine shrimp farms in Haiti. However, there appears to be at least two ventures that are in the final stages of financing. One activity has selected a site of about 3,000 ha in the Artibonite area near Grande Saline. Construction is scheduled to begin in 1989.

According to government officials and representatives of USAID and FAO, there are approximately 8,000 ha of land in Haiti that may be appropriate for the development of shrimp farming.

Freshwater Prawn Farming

One *Macrobrachium* farm has been developed in Haiti. The firm, Aquiculture de Nippes, is operating approximately 10 ha on the southern peninsula, reportedly producing approximately 450 kg of prawns/week. A small hatchery supplies postlarvae to the farm. The firm plans to expand and also to supply other small producers with postlarvae in the near future.

Dominican Republic

Marine Shrimp Farming

There is great interest in developing aquaculture in the Dominican Republic. Marine shrimp farming activities have been concentrated in the area of Montecristi in the north near the border of Haiti.

One of the earliest efforts of shrimp farming is Isabella Acuacultura. The company operates about 80 ha of 1 ha ponds and associated nursery ponds. The primary species being produced is *Penaeus monodon*. However, some *P. schmitti* and *P. vannamei* have also been grown. The ponds, operated by a Taiwanese company, are replicates of the system used in southern Taiwan.

Current production from the farm ranges from 13-18,000 kg/mo., with shrimp being sold as tails and heads-on. Sizes range from 16/20 to 41/50 tails per pound. Isabella Acuacultura operates a full-scale hatchery and a shrimp packing facility. The hatchery produces enough postlarvae to supply the farm. However, it has the capacity to produce 2 million postlarvae/mo. of *P. monodon*. Until recently, most of the shrimp produced has been sold in Santo Domingo where the price is substantially higher than in the United States. However, in 1989 exports to the U.S. will be started.

Another venture, Industria Pesquera Marien, S.A., was started in the Montecristi area. This company had begun the construction of a farm near Manzanillo in 1984. To date, there are 18 nursery ponds of about 0.25-ha in size and 5 growout ponds of about 12 ha each. A seawater supply system has been completed and a small hatchery was constructed in the town of Manzanillo. A few crops were produced approximately 2 years ago. The farm is currently for sale. Several groups are interested in acquiring the property.

The Montecristi area has substantial land for aquaculture, with estimates of as much as 7,000 ha available (NOAA 1988). It appears that more than 10 groups have plans, at various stages of development, to build shrimp farms in this region.

The only other marine shrimp hatchery in the Dominican Republic is west of the Santa Domingo airport at Boca Chica. The hatchery operated by Cultura Mar Caribe (CMC) produces postlarvae of *P. vannamei* for sale to other farms.

The CMC personnel have been stocking and managing a small *P. vannamei* farm at Azua, just east of Santa Domingo. The farm has a total of 5 ha of ponds, ranging in size from 0.5 to 0.8-ha. The ponds have been stocked with postlarvae and harvested after 90 days with yields of 2,042 kg/ha. Shrimp averaging 13 grams heads-on were sold for $8.82/kg on the local market. Recent information indicates that the hatchery sustained damage from Hurricane Gilbert and is now closed.

At present, only two penaeid shrimp farms, Isabella Acuacultura and the small farm in Azua supplied by Cultura Mar Caribe, are in operation in the Dominican Republic. In 1986, penaeid shrimp accounted for only 35 MT of production. In 1987, the total had approached 1,000 MT, primarily supplied by Isabella Acuacultura.

Freshwater Prawn Farming

Freshwater prawn farming in the Dominican Republic is reported to be well developed. Most farms are concentrated in the area of Bayaguana, to the northeast of Santo Domingo (NOAA 1988).

In total, there are more than 900 ha of ponds, operated by 28 farmers, dedicated to the cultivation of *Macrobrachium* (Weidner and Niemeier 1988). Approximately 280 MT of *Macrobrachium* were produced in the Dominican Republic in 1986. There are 12 hatcheries producing *Macrobrachium* postlarvae.

A research station exists at Universidad del Este in the town of San Pedro de Macoris. A small hatchery and research ponds of a few ha have been constructed on the site.

Puerto Rico

The potential of shrimp farming in Puerto Rico has been an area of discussion for some time.

Marine Shrimp Farming

Marine shrimp farming is still an activity for the future in Puerto Rico. There is only one activity with any shrimp, Seafarms of Puerto Rico, which has approximately 0.1-ha of plastic lined ponds stocked with imported postlarvae of *P. vannamei*.

Ten other groups have proposed to build shrimp farms in Puerto Rico. Some of these have been working for as long as five years to obtain permits and develop financing. Three companies appear to be close to startup. Eureka Marine has almost completed construction of approximately 24.4 ha of ponds on a 39 ha site near Dorado on the north coast, some six years from the time the site was selected. They plan to grow *P. monodon* in an intensive system that resembles that used in the south of Taiwan. Camarones de Puerto Rico plans to develop a farm on 170 ha of land in the south near Ponce. The developers project annual production of 816 metric tons after 5 years. Another firm, Camarones del Encanto, is developing a farm of approximately 40 ha near Guayama on the south coast and is expected to begin construction this summer.

With respect to the potential for marine shrimp farming, Aquacultural Concepts estimate that there may be about 1,500 ha of land that may be physically suitable. However, conflicts over land use and problems acquiring a site and obtaining permits present constraints to the development of aquaculture in Puerto Rico at this time.

Freshwater Prawn Farming

Aquaculture Enterprises Inc. operates Sabana Grande Prawn Farm at Sabana Grande, Puerto Rico. The farm, which is one of the largest *Macrobrachium* farms in the world today, is a fully integrated business. A recirculating hatchery with a capacity of 3 million postlarvae/mo. supplies the needs of the farm. There are 90 ponds of approximately 0.5-ha each plus 7 small nursery ponds. The total area of ponds is 48 ha but some ponds have been completed recently and are not yet in full production. The monthly production of prawns from the farm had increased to 9,500 *kg* by May 1989. The average production rate in May 1989 was 2,375 *kg* /ha/year, but is expected to increase to more than 3,500 *kg* /ha/year at full production.

The *Macrobrachium* harvested are sold as whole prawns, with sizes ranging from 23 to 90 grams (20 to 5 per pound). Approximately 30% are shipped to markets in New York, with an additional 25% marketed elsewhere in the United States. Of the remaining 45%, 25% is sold at the farmgate and the rest is sold at restaurants and food stores throughout Puerto Rico (Glude pers. comm.).

In addition to Sabana Grande Prawn Farm, the Department of Marine Sciences of the University of Puerto Rico operates a small experimental hatchery and research ponds at its field station in Lajas. Research is primarily on the polyculture of *Macrobrachium* and tilapia. The Puerto Rican Community Foundation is currently funding a UPR-DMS semi-commercial scale project involving the polyculture of these species.

THE LEEWARD ISLANDS

The shrimp farming activities in the Leeward Islands are limited to two locations where small scale production is taking place: St. Kitts and Antigua.

St. Kitts

In St. Kitts, the Leeward Islands Shrimp Company (LISCO), operated by Nigel Bower, has a total of almost 3.0-ha of ponds producing *P. vannamei*. The farm consists of seven 0.1-ha ponds, one 1.4-ha pond and one 0.4-ha pond. Average production is about 1,362 kg/ha/year. Shrimp produced range in size from 16-24 g. All of the shrimp are marketed on the island, but yields are insufficient to meet the local demand. LISCO operates its own hatchery with a capacity of one million postlarvae/mo.. Postlarvae produced in this hatchery also supply a farm on Antigua and other facilities in the Caribbean.

Antigua

In 1985, Antigua Shrimpery Ltd. constructed a hatchery and three 1.6-ha production ponds on Antigua (Bower pers. com.). The company had financial difficulties and was purchased by East Caribbean Flour Mills. The farm is presently managed by Nigel Bower of the Leeward Island Shrimp Company. The present plan involves stocking the ponds from the St. Kitts hatchery and initiating hatchery activities in the Antigua hatchery over a period of time.

THE WINDWARD ISLANDS

Dominica

Two private sector groups are involved in crustacean culture: a Taiwanese group is rearing *Macrobrachium* and a U.S. company is rearing the Australian freshwater crawfish (*Cherax tenuimanus*) (NOAA 1988; Rakocy and Hargreaves 1986).

The government of Dominica has contracted with the government of Taiwan for technical assistance to develop freshwater prawn farming (NOAA 1988).

St. Lucia

A hatchery to produce *Macrobrachium* postlarvae has been established as a joint effort of the government of Taiwan and St. Lucia (Rakocy and Hargreaves 1986).

St. Vincent

St. Vincent has a limited *Macrobrachium* industry. Several 2,000 m^2 ponds have been built. These are stocked with *Macrobrachium* postlarvae at densities of 35,000/ha. After 4 months, prawns weighing as much as 30 g can be selected by seining (Rakocy and Hargreaves 1986).

Grenada

The OAS has sponsored a pilot activity with the University. A 0.2-ha pond was constructed and stocked with *Macrobrachium* postlarvae in 1986. A harvest in 1987 stimulated interest in aquaculture on the island. There is presently one person who is attempting to establish a semi-intensive penaeid shrimp activity in Grenada.

Trinidad & Tobago

Marine Shrimp Farming

As with many areas of the Caribbean, shrimp farming in Trinidad has been under consideration for some time. In 1984, Montano Shrimp Farms, Ltd., assisted by Aquacultural Concepts, a consulting firm based in Puerto Rico, began to develop an intensive approach to the farming of *P. monodon*. The plan to develop a concession of 400 ha was approved by the government, with the proviso that a pilot project be established to demonstrate the technology. A 1.6-ha facility, consisting of eight 0.1-ha, cement-walled ponds and one earthen pond of 0.4-ha was built and stocked with *P. monodon* from Hawaii. The pilot study was completed in 1987: however, to date, the company has been unable to resolve issues related to lease and equity with the government.

There is great potential for shrimp farming in Trinidad. There are more than 15,000 ha of flat land adjacent to the sea that might be developed into marine shrimp farms. Inland sites of similar size are potential locations for freshwater aquaculture.

Freshwater Prawn Farming

The Institute of Marine Affairs operates a small experimental hatchery and has four demonstration ponds for the production of *Macrobrachium*. Several farmers are considering constructing ponds for *Macrobrachium* culture.

THE FRENCH WEST INDIES

Guadeloupe

On Guadeloupe there is a *Macrobrachium* hatchery operated by a co-operative of prawn farmers, Sica Guadeloupeene d'Aquaculture (NOAA 1988). The hatchery, built in 1984, has an annual capacity of 25-30 million postlarvae. This facility has been producing postlarvae for sale in the region. In 1985 and 1986, 6.5 million postlarvae were produced and sold to Puerto Rico, Dominica, and Grenada.

Guadeloupe has 19 *Macrobrachium* farms, totaling about 60 ha in production. The farms range in size from 0.2 to 12 ha. In 1987, the farms produced 65 MT of *Macrobrachium* (Linden 1987). The average yield is about 1.5 MT/ha/yr (NOAA 1988). To date, the prawns produced in this country have been sold whole at the farmgate or to local markets for $22-$26/kg fresh or frozen.

A large hatchery has been developed jointly by France Aquaculture and the French Institute of Research for the Exploitation of the Sea (IFREMER). The hatchery has the capacity to produce 15 million PL's per year. This facility produced 9 million PL's in 1986 (NOAA 1988).

Martinique

IFREMER has also been active in developing *Macrobrachium* farming in Martinique. In Martinique, there are currently more than 80 farmers growing prawns in about 200 ha of ponds. Production is estimated at 50 MT/yr (Weidner and Niemeier 1988). Postlarvae are supplied by a cooperative hatchery which is owned by the farmers and has the capacity to produce 12 million postlarvae/yr. A government survey indicates that there may be as much as 400 ha of land suitable for the development of prawn ponds on Martinique.

CONCLUSIONS

Compared to many other parts of the world, shrimp farming in the Caribbean is under developed. In 1988, only 1,635 MT was produced in the Caribbean islands. This represents only 0.15% of the total shrimp consumed in the U.S. in 1987 and 0.2% of total imports (1.7 million M.T and 1.3 million MT, respectively). If we discount the production in Cuba, which is not well documented, the total production in the Caribbean in 1988 was less than 0.12% of the shrimp imported into the United States during that period. By comparison, Ecuador, the largest supplier of shrimp to the U.S. market, contributed 70,000 MT in 1988. This is almost 43 times the amount produced from the Caribbean Islands. There is no single reason for this situation. Some of the factors that may have contributed to this situation are as follows:

1) Lack of Suitable Species in the Caribbean

In contrast to the Pacific coast of Latin America and areas of Southeast Asia where gravid females and postlarvae are abundantly available, there is a lack of indigenous species and supplies of postlarvae to support shrimp farming in the Caribbean (Stoner 1988). The development of shrimp farming will require reliable and economic supplies of postlarvae to sustain production.

2) More Advantageous Conditions in Other Locations

Large tracts of land and opportunities for expansion have previously been available in other locations of the world. However, many of these areas have been developed to their potential and attention has recently been directed to the Caribbean. Also, the development and verification of technology for intensive culture now makes it possible to establish viable shrimp farms in smaller tracts of land.

3) Government bureaucracy

At present, many of the governments in the Caribbean have no established procedures for timely approval of permits and licenses needed to develop aquaculture activities. In many cases, this has led investors with technology and financing to look in other areas.

4) Skepticism

Some governments in the Caribbean have been skeptical or indifferent to the potential of shrimp farming in their countries. In some instances, requests for sites or assistance are not taken seriously.

5) Competitiveness

In some locations, limitations on land and high production costs make shrimp farming unprofitable or non-competitive.

We believe that this year marks a change for the Caribbean. In locations such as Haiti, Jamaica, the Dominican Republic, Puerto Rico, Trinidad, where appropriate resources are available, an appreciation for the potential for shrimp aquaculture is developing. Some significant activities are now underway in the Caribbean. The countries in the region have resources that are sufficient to develop significant shrimp farming. It behooves the countries to examine and pool their resources to develop shrimp farming to its full potential as a regional activity.

LITERATURE CITED

Linden, E. 1987. The development of freshwater prawn farming in Guadeloupe (French West Indies). The Caribbean Aquaculturist 3(3): 10.

National Oceanic and Atmospheric Administration, Washington DC, US Department of Commerce. 1988. Latin American shrimp culture industry 1986-90. pp. 38-48.

Rakocy, J.E. and J. Hargreaves. 1986. Assessment of aquaculture in the Eastern Caribbean. Publication of the College of the Virgin Islands, Eastern Caribbean Center, August 1986. pp. 6-17

Rosenberry, R., editor. 1989. Aquaculture Digest. A monthly report on fish and shellfish farming. p. 18. March 1989

Stoner, A.W. 1988. A nursery ground for four tropical *Penaeus* species: Laguna Joyuda, Puerto Rico. Marine Ecology - Progress Series 42:133-141.

Weidner, D. and P. Niemeier. 1988. Shrimp aquaculture in Latin America. Appendix II. Pages 109-171 *in* Aquaculture and capture fisheries: impacts in US seafood markets. US Department of Commerce, NOAA, NMFS., Washington, D.C.

Wint, S.M.E. 1988. Aquaculture development in Jamaica. Paper prepared for the 41st Annual Meeting of the Gulf and Caribbean Fisheries Institute, St. Thomas, USVI, November 6-11 1988.

STATUS AND HISTORY OF MARINE AND FRESHWATER SHRIMP FARMING IN SOUTH CAROLINA AND FLORIDA

J.Stephen Hopkins

INTRODUCTION

The first documented attempts at marine shrimp farming in North America and the Caribbean took place in South Carolina and Florida. Since that time, there has been sporadic shrimp farming activity with the present level of interest relatively high. The history of marine shrimp farming in South Carolina and Florida parallels the development and intensification of the marine shrimp farming technology for the Western World in general. Many of the major advances in marine shrimp culture have been the result of work done in South Carolina and Florida and many recognized experts in the field began their careers in these states. In addition, there have been numerous research and commercial projects to produce freshwater shrimp. Unfortunately, there has never been a large-scale marine or freshwater shrimp farming operation in South Carolina or Florida which could be considered an economic success. It remains to be seen whether current efforts to increase profitability through intensification of pond production can foster a truly viable shrimp farming industry in the area.

MARINE SHRIMP

The history of marine shrimp farming in South Carolina and Florida can be characterized by several stages of development which, in general, correspond to technological advances. In the earliest days of shrimp farming, existing impounded wetlands were stocked through recruitment of wild postlarvae from adjacent estuaries. Later, it became possible to stock postlarvae that had been produced in a hatchery. The availability of postlarvae as a marketable commodity simplified the process of getting started in the shrimp farming business. Finally, intensification of pond production has made shrimp farming in South Carolina and Florida more cost effective and may lead to development of an industry which is truly viable in an economic sense.

Tidal Impoundment Recruitment Method

Shrimp can be effectively drawn into an impounded wetland by capitalizing on their natural instinct to move into the upper reaches of the estuary. This practice is particularly effective in areas with high tidal amplitudes. Flap-gate water control structures, originally developed for impoundment culturing of rice, are ideally suited to recruiting shrimp postlarvae into the pond.

The practice of tidal impoundment recruitment is crude as compared to currently available cultivation methods. The unpredictability of postlarval recruitment, predator control problems, the large land requirements, and the labor-intensive harvesting techniques makes it unlikely that this form of shrimp farming will ever expand greatly. However, the recruitment method demonstrated that it is indeed possible to produce marine shrimp outside of their natural ecosystem. This, in turn, prompted efforts to develop hatchery supplies of postlarvae which have eventually led to more intensive production systems.

Lunz (1951) reported on recruitment of indigenous penaeids into tidal impoundments along the coast of South Carolina. Lunz and Bearden (1963) further refined management techniques and made the information available to owners of impounded wetlands. Whetstone (pers.comm.) estimates that there are 4,000 to 5,000 ha of tidal impoundments in South Carolina alone suitable for extensive shrimp production.

Notable among the numerous tidal impoundments in South Carolina which are occasionally used to recruit and harvest marine shrimp is Annandale Plantation. Annandale Plantation's initial experience in shrimp farming was in conjunction with the Palmetto Aquaculture group but, more recently, Annandale has harvested extensively produced crops of recruited shrimp on its own. Other Georgetown County impoundment owners such as Kinlough Plantation and Esterville Plantation have also been successful in recruiting and harvesting shrimp. Some particularly impressive harvests were noted in the 1960's at Seaside Farms in Charleston County. Whetstone, et al. (1988) summarized recent work on extenisve shrimp culture in South Carolina, and illustrated the types of yields possible.

Whether or not these impoundments are used for extensive shrimp production in a given year depends on a number of factors including the strength of the wild *Penaeus setiferus* year class, salinity levels and the availability of labor for pond management. The South Carolina population of *P. setiferus* varies greatly from year to year in response to the winter survival of the spawning stock. Certain impoundments are also adversely affected by higher than normal amounts of rainfall, which moves the estuarine saltwater wedge seaward and inhibits postlarval migration to the impoundment gate. In summary, as extensive shrimp production based upon recruitment stocking is an unpredictable endeavor, many impoundment owners are reluctant to expend the effort necessary to effectively manage the pond.

Hatchery Technology Development and Its Impact

The shrimp farming industry could not have matured into a viable industry world-wide were it not for the development of effective hatchery technologies. The availability of hatchery-reared postlarvae fostered the process of intensifying grow-out production and increased the attractiveness of marine shrimp farming as a reasonable corporate-level enterprize.

With the "Galveston Method" of predictable and cost effective hatchery techniques (see Chamberlain 1991), industry development, through public and private research and production projects, shifted to Florida in the late 1960's. Notable among the farms of this era were the Turkey Point project, Marifarms, and the Crystal River project.

The Turkey Point project operated from 1967 through 1972 and was funded by Armor and United Brands through the University of Miami. At Turkey Point, larval rearing techniques were refined, marine shrimp species evaluated for production and diets developed. The project attempted to utilize the waste heat from a Florida Power and Light nuclear power plant to overcome the climatically restricted growout season. Unfortunately, the presence of tritium in the recirculating water precluded commercial use of the product and financial backing was withdrawn.

Perhaps the greatest contribution of the Turkey Point project to the present day shrimp farming industry was the training of personnel who participated in the project. Many of these individuals went on to establish profitable enterprises in Central and South America and remain leaders in the field today.

One of the largest shrimp farming investments ever made was the Marifarms project which operated in Florida during this time. Marifarms operated a 1000 ha netted bay enclosure plus two ponds with a total area of 255 ha. The growout units, being quite large, were plagued with predator control problems and storm damage to the net enclosure. Marifarms' assets and lease rights were sold to the present-day Continental Fisheries Ltd. in 1982. Marifarms personnel went on to undertake a joint venture with an Ecuadorian hatchery.

Important contributions of the Marifarms and Continental Fisheries operations have been the evaluation of several native and non-indigenous species for production and the refinement of large-scale hatchery systems capable of producing up to 200 million *P. setiferus* postlarvae during the three-month natural spawning season.

Another important shrimp farming project was conducted at Crystal River, Florida from 1972 until 1982. The Crystal River project was associated with a fossil fuel power plant and funded by the Ralston Purina Company. Here, the Pacific Coast species, *P. stylirostris* and *P. vannamei,* were utilized. This project was the pilot for Purina's successful operation in Panama which was more recently sold to the Granada Corporation.

Perhaps the most significant contribution that the Crystal River project made to present-day shrimp farming technology was the development of commercially viable controlled maturation systems. The controlled maturation technology has made a major impact on the development of farms in South Carolina and Florida. Controlled maturation makes possible the production of postlarvae of non-native species which may have more desirable culture traits than native species. Controlled maturation also allows the hatchery to produce postlarvae before the natural spawning season so as to take full advantage of the limited growing season.

A fourth penaeid growout venture to operate in Florida was Sea Farms. Sea Farms utilized the turtle corrals at Key West and contemplated another growout site at Tarpon Belly Key. Sea Farms went on to establish a profitable enterprise in Honduras. This group is now operating farms in various parts of Latin America through the Miami-based Shrimp Culture Inc. Sea Farms also established a hatchery on Summerland Key.

Postlarvae as a Commodity

The necessity of a reliable supply of hatchery-reared postlarvae initially posed certain restrictions on the siting and size of shrimp farms. Requirements for oceanic quality water for larval rearing dictate that a hatchery be located on or near expensive beach-front property. Beach front property is in high demand for a number of other uses and is, thus, often very expensive and very limited in availability. Property with oceanic quality water is not necessary for grow-out facilities which can operate effectively using estuarine water adjacent to less expensive land.

Economies of scale, particularly with respect to justifying the employment of highly trained hatchery personnel, dictate that a hatchery facility be fairly large. Thus, expansive grow-out production facilities have to be available to accept hatchery output. Small scale or inexpensive pilot level combined hatchery-growout operations are thus impractical.

This limitation can be circumvented by the sale of postlarvae from large hatcheries to smaller grow-out operations. Postlarvae may be purchased from hatcheries which are either separate business entities or part of a larger integrated farm. Large integrated farms can oversize their hatchery capacity in the design phase to provide a margin of safety should there be hatchery production problems, and thus generate extra revenues when hatchery production is high.

For the past five years, the Continental Fisheries Ltd. hatchery discussed above has been a major supplier of postlarvae for shrimp farms in South Carolina and Texas as well as supplying operations in Latin America. After the Marifarms operation was purchased by Continental Fisheries in 1982, the two growout ponds operated for two more years. Grow-out operations were finally dropped completely due to the problems noted above and, beginning in 1984, the company concentrated on hatchery production and postlarval sales. Prior to 1985, Continental Fisheries produced postlarval *P. setiferus* from wild adults captured by trawler off the Florida coast. Maturation units were added to facilitate controlled reproduction of *P. vannamei* in response to the demand for this species. During 1985-1987, the hatchery operated year-round to supply U.S. and Latin American markets. In 1988 and again in 1989, the hatchery operated only seasonally to supply farms in the U.S.

Sea Farm, the predecessor of Shrimp Culture, Inc., has had a hatchery facility in and out of operation at Summerland Key since 1971. For a time, the Summerland Key facility was leased and known as Florida Keys Aquaculture. Shrimp Culture Inc. has recently finished the permit process for an expansion at the site and is resuming production as a corporate cost center supplying postlarvae to their Honduras grow-out facilities.

Open commercial supplies of postlarvae and cost effective techniques for shipping them has facilitated the creation of semi-intensive shrimp farms in South Carolina (Hopkins et al. 1987). The first such endeavor was Palmetto Aquaculture. Palmetto Aquaculture's approach has been to minimize construction and operating costs while slowly intensifying the production strategy. For growout, the company leases idle impounded wetlands on a share-crop basis. The company began operation in 1981 stocking *P. stylirostris* at Annandale Plantation and has since utilized numerous existing impoundments in South Carolina. Hatchery reared *P. setiferus* were stocked between 1982 and 1984 before production was switched to *P. vannamei*. Palmetto Aquaculture began operating at a modest profit in 1987. The technology they have developed is profitable provided survival is above 50%.

In 1983-84, Plantation Seafarms was constructed in an impounded wetland on Edisto Island, South Carolina. After two years of semi-intensive operation, Plantation Seafarms expanded to a new highland site. Plantation Seafarms was financially restructured and is now called Edisto Shrimp Co. *P. setiferus* was cultivated for several years before switching to *P. vannamei* with some recent diversification into shellfish and finfish. Edisto Shrimp Co. has the largest annual production of farm-raised shrimp in the history of the area. The company has, at several points, seriously considered the construction of an integrated hatchery as well.

The original Plantation Seafarms site was sold in 1986 and is now operated as Sand Creek Shrimp Farm. The facility has been upgraded and the management inputs increased. The size of the individual production units was decreased, aeration equipment installed, and feed storage and other buildings added. The farm is also diversifying into shellfish biculture and bait shrimp. Sand Creek was the first owner-operated (as opposed to corporate-owned) semi-intensive farm to develop in South Carolina during this period.

Intensification of Grow-Out

Plantation Seafarms found that highland ponds can generally be managed much more effectively than the typical tidal impoundment. Highland areas are less prone to problems with acidic soils, and can easily be drained and dried to facilitate removal or oxidation of accumulated organic material. Many feel that the ease of management and potential for increased production more than justifies the higher land purchase cost of high land versus tidal impoundments.

As semi-intensive shrimp farms were developing in South Carolina, the South Carolina Marine Resources Division built the Waddell Mariculture Center to provide research and development support for shrimp farming as well as other mariculture opportunities. Research has determined that the production levels from highland shrimp ponds can be increased dramatically through the use of supplemental aeration. At the Waddell Mariculture Center, production levels in excess of 20,000 kg/ha have been demonstrated in rectangular ponds originally constructed for semi-intensive use. More importantly, economic analysis (Rhodes 1991) has indicated that intensification can result in higher overall returns for a new project.

Based upon this information, new shrimp farms built in South Carolina are designed to take advantage of intensive production technologies. The total number of farms has increased dramatically over the past several years (Table 1). Newer ponds are generally smaller and thus more manageable at intensive production levels.

Table 1. Estimated production of cultivated marine shrimp in South Carolina 1980 to 1988.

	RECRUITMENT STOCKED				HATCHERY STOCKED			
Year	No. Farms (n)	Total Area (ha)	Total Prod. (mt)	Yield/ Area * (kg/ha)	No. Farms (n)	Total Area (ha)	Total Prod. (mt)	Yield/ Area * (kg/ha)
1980	2	506.0	4.0	7.9	0	0	0	0
1981	2	607.0	4.5	7.4	1	20.2	0.1	5.0
1982	3	810.0	6.3	7.8	1	394.6	2.9	7.3
1983	3	972.0	8.8	9.1	1	111.3	2.5	22.5
1984	5	850.0	4.3	5.0	2	32.3	12.7	393.2
1985	5	850.0	3.8	4.5	3	49.6	24.7	498.0
1986	5	800.0	12.0	15.0	4	36.2	68.4	1889.5
1987	3	854.0	6.8	8.0	7	74.3	166.2	2236.9
1988	4	806.5	2.3	2.9	11	107.0	245.1	2290.7

* kg/ha = whole weight per crop, one crop per year.

For example, Richardson Plantation began producing shrimp in 1986. This large family trust diversified into shrimp farming with a pilot operation which has grown slightly each year as various production intensities and pond configurations have been tested. The management has made dramatic increases in yield each year and in 1988 broke all western-world records for commercial pond growout with 15,000 kg/ha/crop whole weight in a 0.3 ha pond. Another year of pilot operation is anticipated before expansion to full scale.

The Toogoodoo Shrimp Farm began operation in 1987. This farm operated for two years in an impoundment pond. In 1989, the farm expanded through the addition of two adjacent highland ponds. In addition to these ponds, the owner-operator is involved in the management of several impoundment ponds at remote sites.

As Plantation Seafarms was being financially reorganized in 1986, they dropped plans for a small hatchery which was under construction on a leased site. The owners of this leased site used the unfinished infrastructure as the foundation for Edisto Mariculture in 1987. Edisto Mariculture has managed an impoundment for shrimp at nearby Rabbit Point, and has intensively cultured shrimp in what were originally intended to be hatchery settling ponds. In addition, Edisto Mariculture has completed the permit process for a hatchery operation.

In 1988, the number of shrimp farms in South Carolina increased dramatically with the addition of seven new farms. One of these, Sea Fare Inc., is a corporate-level farm which leased a lowland site and renovated it for semi-intensive production. While all other present-day farms started with or have switched to paddlewheels as the source of aeration, Sea Fare is developing an intensive pond technology which utilizes diffused air. The company is seeking a permit for the creation of numerous small ponds and water distribution canals inside what is now the larger pond.

Five of the seven farms which began operation in 1988 experienced construction or stocking delays which resulted in reduced crops their first year. However, by moving shrimp though outlets other than the commodity market, these farms were able to provide reasonable cash flows and refine management skills. Taylor Creek Shrimp Co. was able to complete construction in time for a near normal season. In addition, this company has the equipment necessary to construct ponds and has built facilities for two other farms. Taylor Creek and Sea Fare both conducted pond-side sales which bring a substantial premium over the commodity price.

Perhaps the most inland shrimp farm ever built in the area is Tullifinny River Co. Like Richardson Plantation, Tullifinny River Co. operates shrimp ponds as part of a large landholding with diversified agriculture components. This farm draws water from a brackish source which may occasionally be nearly freshwater. Delays in installation of the water distribution system necessitated low water exchange and reduced production goals for the 1988 season. The management did not put ponds in production in 1989 due to another commitment.

Huspa Plantation shrimp farm began operations late in the 1988 production season. All of the product was sold as live bait at premium prices.

The farm has installed a greenhouse covered nursery to head-start shrimp before the outdoor growout season. Further production of live shrimp for fish bait is expected due to initial success with this approach.

CLP Trading Co. also completed pond construction and began operations late in the 1988 season. Here again, small shrimp were produced due to the loss of half of the production season so shrimp were sold either as bait or as fresh head-on product for the oriental restaurant trade. Like Huspa Plantation and Taylor Creek Shrimp Co., CLP Trading Co. is a family operated business. However, CLP has investments in shrimp farms in the Philippines and is one of the few farms whose managers have had prior experience in shrimp culture.

Finally, additional impoundment ponds were put into extensive production using hatchery seed in 1988 and 1989. Management personnel at Spring Island, a sizable land holding, is for the second time stocking shrimp in one of its impoundments with encouraging results. At Oak Grove Plantation, personnel from the former Plantation Seafarms stocked an impoundment in 1989.

In 1989, Atlantec Seafarm, a new corporate level farm was constructed and began operations. This 4 ha pilot farm plans a large expansion if initial results are encouraging and permits for the expansion can be obtained.

In recent years, most new farms have been small and owner-operated. This may reflect the limited availability of venture capital for corporate-level farms due to the risk factors involved or the general state of the economy. Several moderately large farms, such as Atlantec Seafarm and RPI International, have been designed but have not yet solicited all the needed financing.

FRESHWATER PRAWNS

The South Carolina to Florida region also has a history of research on, and commercial production of freshwater prawns, *Macrobrachium rosenbergii.* Research and development efforts have been conducted by both South Carolina and Florida state agencies beginning in 1972. Integrated commercial hatchery and growout facilities were built in Florida as well as grow-out operations in South Carolina. However, there is presently very little freshwater prawn farming activity in the area.

Research on *Macrobrachium* sp. farming was initiated by the Florida Department of Natural Resources and the South Carolina Marine Resources Division in the early 1970's (Dugan et al. 1975, Sandifer and Smith 1974). In both cases, initial work involved indigenous species such as *M. acanthurus*, *M. ohione* and *M. carcinus*. After some initial larval rearing and tank growout trials, both agencies moved the emphasis of their work to the Malaysian species *M. rosenbergii*. This species exhibits rapid growth under the proper conditions and is only moderately aggressive.

During the 1970's, several universities and private foundations also became involved in *Macrobrachium* sp. research. Florida Atlantic University conducted research projects in tanks and ponds on campus. The Jacksonville University and the University of Miami also conducted laboratory and tank growout studies. In South Carolina, *Macrobrachium* sp. engineering and physiology studies were conducted at Clemson University and the Medical University of S.C. The Harbor Branch Foundation in Florida conducted pond growout and tank studies.

Several large corporations became involved in pilot *Macrobrachium* sp. farming projects. Notable among these were operations by the Weyerhauser Company in Florida and the Alexander-Dawson Corp. in South Carolina.

Numerous small entrepreneurial operations to farm freshwater prawns developed in each state as well. Notable among these in Florida were Farm Fresh Shrimp which operated a hatchery and growout facilities. A larger facility was operated in the West Palm Beach area where about eight hectares were put into growout production. The operation which most nearly approached success may have been Florida Aquaculture which built a series of ponds for *Macrobrachium* sp. production before switching to tropical fish.

In South Carolina, several individuals converted small farm ponds or built ponds for *Macrobrachium* sp. growout during the early 1980's. At its peak, there were nine freshwater prawn farmers in the state with a total of 3.25 ha of water area. Pond size ranged from 0.04 to 0.4 ha and most operations had but one pond. Production from these operations was highly dependent upon the degree of water control possible and the amount of time available for pond management (Smith et al. 1982). The maximum production recorded was 1,900 kg/ha, although the average was more in the order of 600 to 700 kg/ha.

Introduction of *Macrobrachium* sp. farming into the South Carolina and Florida area was a success in a technical sense but not financially. The inability to achieve acceptable growth rates at high density made profitable cultivation in an area with high capital construction costs impossible. The problem was further compounded by the restrictive climate.

CURRENT STATUS

At present, nearly all of the shrimp cultured in South Carolina and Florida is the Pacific species *Penaeus vannamei*. However, other penaeids and some *Macrobrachium rosenbergii* are produced on an occasional basis.

In 1989, there are a total of twelve semi-intensive or intensive pond shrimp farms in South Carolina. These farms have a combined water area of 113 ha. Four other farms have 800 ha of impounded wetlands for extensive production using recruited postlarvae (Table 1).

All of the growout farms in South Carolina are geared to produce but one crop per year with the possible exception of bait producers. The climatic restriction to one crop per year and rising land costs have encouraged intensification of production. However, the intensification process has an impact on harvest size (Sandifer et al. 1988) as there is some degree of growth suppression at higher densities. Therefore, the harvest size of most shrimp produced in South Carolina at present is 14 to 19 g. Sales of bait shrimp commence as soon as the shrimp reach a size of 4 g and the sales continue through the season as the shrimp grow.

To date, most farm-raised shrimp produced in South Carolina have been sold through the same docks which handle the fishery product. The numerous middlemen between the farmer and the consumer results in pond-side prices which are below the commodity prices as published by shrimp marketing publications. However, in recent years, there has been a push to develop specialty markets such as those for bait shrimp, individually quick frozen product, head-on sales, pond-side sales to consumers and fresh sales to restaurants and retail outlets.

In 1988, the farm-raised shrimp crop in South Carolina was roughly valued at $1.1 million. The amount of capital investment to date is unknown but is probably in excess of $3 million.

Very little freshwater prawn farming continues in either South Carolina or Florida. One farm in South Carolina still produces a pond of *Macrobrachium* sp. on an occasional basis. This effort is limited by the availability of postlarvae at a reasonable price. There was no freshwater prawn production in 1988 or 1989.

INDUSTRY SUPPORT

The South Carolina/Florida area has developed considerable support services for the shrimp farming industry. These support services serve both the South Carolina/Florida area and shrimp farms located in Latin America and Texas. Most of these support services are located in Florida, as Florida has easy access to the large shrimp farming industries in Latin America. Support services include equipment manufacturing and sales, postlarval suppliers, product brokerage and processing, training programs, research and development support, extension services and consulting services.

Much of the generalized equipment used in Latin American shrimp farming is manufactured in the U.S. and purchased and shipped through Miami. In addition, there are companies which manufacture specialized shrimp farming equipment in Florida. Shrimp pond aerators and shrimp feed are manufactured in South Carolina.

Until recently, all shrimp feed used in South Carolina was shipped in from out-of-state manufacturers. Public agencies have attempted to educate South Carolina feed manufacturers about the demand for aquaculture feeds and in 1989 the first locally produced shrimp ration became available. Local mills have a competitive advantage of lower shipping costs to South Carolina farms but it is yet to be seen whether the feeds they produce perform as well as those of more established aquaculture feed manufacturers.

As noted above, hatcheries in Florida supply postlarvae to growout farms in South Carolina and Texas as well as Latin American. However, in recent years the shrimp farms in South Carolina and Texas have had to rely more and more upon Latin American hatcheries as the production of the Florida hatchery has not been able to keep up with the domestic demand during the short pond stocking window in the spring.

Much of the shrimp farmed in Latin America is destined for U.S. markets. Since Miami is a major port of entry from Latin America, many shrimp brokerage firms are located in this area. Florida, with a long history of processing much of the shrimp captured in the South Atlantic, Gulf and Caribbean fisheries, also processes cultured product in these same facilities.

As noted above, the research and development work and early commercial ventures in Florida have produced many highly skilled shrimp farmers. In recent years, specialized shrimp farming training programs have been conducted by firms based in Florida and South Carolina.

Public agency research and development support for shrimp farmers was, at one time, available in both Florida and South Carolina. The state of Florida no longer provides the commercial shrimp farmer research and development support. In South Carolina, shrimp farming research and development programs are at an all-time high and increasing. South Carolina also has a system of transferring technology to the industry and responding to industry related problems through state extension services.

Finally, there are numerous firms in both Florida and South Carolina which provide consulting services to shrimp farmers world-wide. These services are available for every aspect of the industry from farm design to marketing. Thus, while the production level of farm-raised shrimp in South Carolina and Florida is still relatively small, a large body of expertise is available to support the industry.

PROBLEMS AND PROSPECTS

Small marine shrimp farming industries in Florida and South Carolina have their problems. Some of these problems are largely uncontrollable, such as climatic restrictions and competition from foreign imports. Other problems, such as high feed costs or inability to obtain permits, are resolvable, but would require concerted efforts on the part of government and industry.

A major difficulty in farming shrimp in Florida and South Carolina is the climatically restricted growing season. While this area has ample growing season to produce one crop, the temperate farmer must compete in a market place supplied by tropical farms which produce two or more crops per year. This puts the farmer at an economic disadvantage in terms of direct costs and amortization of investments. The South Carolina or Florida shrimp farmer must seek to overcome the climate disadvantage with advanced technology, proximity to prime markets and the stable political and economic situation that exists in the U.S.

Foreign shrimp producers may have other advantages over the U.S. farmer. In other countries, the cost of certain major operating costs, such as fuel or labor, may be much lower than that of the U.S., allowing the foreign producer to export shrimp to the U.S. at a price below the domestic cost of production.

A significant hindrance to the construction of new marine shrimp farming facilities in the U.S. is the coastal permitting systems. Coastal zone management regulations will not permit a shrimp farm to be built in wetland areas. However, existing impounded wetlands can be maintained in this state. There are few restrictions on pond construction on high ground adjacent to marine waters.

Water quality protection regulations make it extremely difficult to obtain permits to withdraw water from an estuary or ocean beach in Florida. In South Carolina, permits to withdraw water can be obtained provided the pumping structure does not interfere with navigation, the construction does not destroy marshlands and the intake velocities are reduced to minimize entrainment of marine life. In addition, new permits in South Carolina are requiring that zooplankton being filtered at the pond inlet be returned to the estuary.

Both South Carolina and Florida regulate the discharge of water back to the receiving body. It is thought that a permit to discharge water from a shrimp farm back to the marine environment would be extremely difficult to obtain in Florida. In South Carolina, discharge permits can be obtained as long as there is sufficient dilution to prevent the biological load in the effluent from significantly impacting the water quality in the receiving body. Research is underway in South Carolina to develop cost effective water polishing systems for pond effluent, thus increasing the total amount of shrimp that could be farmed along a particular estuarine body.

Land availability may, in the long run, be a significant deterrent to further development of a marine shrimp farming industry. The need for seawater puts the potential shrimp farmer in direct competition with other development interests for waterfront property. The problem is becoming acute in Florida, and to a slightly lesser extent, in South Carolina as well. In South Carolina, for example, the cost of waterfront property in the Edisto Island area has almost doubled in the past five years.

Construction of new farms is also hindered by the availability of capital to finance projects. Many of the options for soliciting financing for a new business are not available to the potential shrimp farmer due to the high perceived risk. When available, financing may be very expensive due to the risk factor which, in turn, further limits potential profitability. Venture capital has become increasingly difficult to find due to changes in the U.S. tax codes.

Shrimp farms in South Carolina are faced with several immediate operational problems. The supply of penaeid postlarvae available to the U.S. farms varies greatly from year to year. The restricted growing season demands that the farmer stock outdoor ponds over a fairly narrow window of opportunity in the spring. Postlarvae supplies were low in the 1989 season, although all farms did eventually obtain sufficient stock. Supplies for the 1990 season are expected to be extremely low, thus plans are underway to develop local hatchery capabilities and locate additional sources of postlarvae out-of-state.

Six commercial penaeid hatcheries have supplied the U.S. market over the past few years. These hatcheries are located in Florida, Texas, Hawaii, Panama and Costa Rica. All except the Florida hatchery (Continental Fisheries Ltd.) are part of integrated growout farms, which take precedence over sales to other farms. The Florida hatchery finds it difficult to operate on a seasonal basis and is unable to export postlarvae during the nine months in which there is no domestic demand. As discussed above, the other Florida hatchery (Sea Culture Inc.) exclusively supplies its own growout ponds in Latin America. The Panama hatcheries are also part of growout farms but, despite their own requirements, the long shipping distance and importation regulations, have been able to supply a large portion of the U.S. market in recent years. However, the political situation in Panama may restrict the export of postlarvae from that country in the future. Other foreign hatcheries interested in selling to the lucrative U.S. postlarvae market are being solicited by growout farmers in South Carolina.

A final problem for the growout farmer in South Carolina is the rising cost of shrimp feed. Feed prices rose dramatically during the 1988 production season in response to drought conditions and a general rise in the cost of all sources of protein. In the 1989 season, feed prices have moderated considerably. This problem is being addressed on two fronts. The efficiency of feeds in terms of the price and growth rate they will sustain is being refined through cooperative work between the feed manufacturers and research facilities. In addition, other feed producers are being encouraged to enter the shrimp feed market which should create competition between the manufacturers and, hopefully, reduce the cost to the farmer.

ACKNOWLEDGMENTS

This paper would not have been possible without the gracious exchange of information provided by many industry and agency representatives including: Bart Baca, David Belanger, David Cannon, Tom Cox, Dana Dunkelberger, Bobby Ellis, Chris Howell, David Janney, Bruce Martin, Gordon Mobley, Bill More, Joe Mountain, Jim Norris, Caesar Pangalangan,

Graham Reeves, Ted Smith, Dick Stansell, Al Stokes, Barbara Sturgill, Marion Stone, Durban Tabb, Frank Taylor, Joe Wannamaker and Jack Whetstone. Special thanks are due George Chamberlain, Chris Howell and especially Nora Bynum who critically reviewed the manuscript and made many useful suggestions.

During the editorial process, the information presented in the original oral presentation at the Annual Meeting of the World Aquaculture Society (Los Angeles, California, February, 1989) was updated and expanded in response to new developments within the industry.

LITERATURE CITED

Chamberlain, G.W. 1991. Status of shrimp farming in Texas. Pages 36-57 *in* P.A. Sandifer, editor. Shrimp culture in North America and the Caribbean. World Aquaculture Society.

Dugan, C.C. 1971. Florida research: freshwater shrimp. American Fish Farmer & World Aquaculture News 2(10):9.

Hopkins, J.S., D.T. Belanger, M.A. Salinger, D.C. Payne, D. Fuze, A.M. Stone, B.G. Ellis, P.A. Sandifer and A.D. Stokes. 1987. The magnitude and potential of the semi-intensive shrimp farming industry in South Carolina. Abstracts of the Annual Meeting World Aquaculture Society, Guayaquil, Ecuador, January 1987.

Lunz, G.R. 1951. A saltwater fish pond. Contributions, Bears Bluff Laboratory, South Carolina 12:1-12.

Lunz, G.R. and C.M. Bearden. 1963. Control of predacious fishes in shrimp farming in South Carolina. Contributions, Bears Bluff Laboratory, No.36, 96pp.

Rhodes, R.J. 1991. U.S. shrimp farms: can they survive. Pages 202-215 *in* P.A. Sandifer, editor. Shrimp culture in North America and the Caribbean. World Aquaculture Society.

Sandifer, P.A. and T.I.J. Smith. 1974. Development of a crustacean mariculture program at South Carolina's Marine Resources Research Institute. Proceedings of the World Mariculture Society 5: 431-439.

Sandifer, P.A., J.S. Hopkins and A.D. Stokes. 1987. Intensive culture potential of *Penaeus vannamei*. Journal of the World Aquaculture Society 18(2): 94-100.

Smith, T.I.J., P.A. Sandifer, W.E. Jenkins, A.D. Stokes and G.Murray. 1982. Pond rearing trials with Malaysian prawns, *Macrobrachium rosenbergii*, by private growers in South Carolina, 1981. Journal of the World Aquaculture Society 13:41-55.

Whetstone, J.M., E.J. Olmi, III and P.A. Sandifer. 1988. Management of existing saltmarsh impoundments in South Carolina for shrimp aquaculture and its implications. Pages 327-338 *in* D.D. Hook et al. editors. The ecology and management of wetlands. Croom Helm, London; Timber Press, Portland, OR.

STATUS OF SHRIMP FARMING IN TEXAS

George W. Chamberlain

BRIEF HISTORY

Research Milestones

Larval Rearing

The origins of penaeid shrimp farming in Texas can be traced to the early 1960's when Harry Cook and coworkers with the National Marine Fisheries Service (NMFS) Laboratory in Galveston, Texas, developed techniques for spawning and larval culture of several Texas penaeids including *Penaeus aztecus*, *P. duorarum*, and *P. setiferus* (Cook and Murphy 1966; Cook 1969; Mock and Murphy 1970). The NMFS technique modified the large-tank, community-culture method pioneered for larval culture of *Penaeus japonicus* by Japan's Dr. Motosaku Fujinaga, now recognized as the "Father of Shrimp Culture" (Hudinaga 1942). Harry Cook credits much of his early success at NMFS to the existing algae culture expertise at the Galveston Laboratory due to ongoing red tide research. Also, non-commercial penaeids, such as *Sicyonia*, *Xiphopenaeus*, and *Trachypenaeus*, were readily available for practice trials through use of local chartered vessels. The Galveston Laboratory continued to refine larval rearing methods throughout the 1970's and served as an important demonstration and training center for aspiring hatchery biologists worldwide. The methods utilized by the NMFS researchers are still widely known as "the Galveston Laboratory Technique" (Mock et al. 1980; Wilkenfeld et al. 1983; and Kuban et al. 1985).

Pond Grow-Out

The early success in larval rearing of shrimp stimulated interest in growing shrimp to market size in ponds. Initial pond trials by NMFS, Texas A&M University System, and Texas Parks and Wildlife Department relied upon postlarval and juvenile *P. aztecus* and *P. setiferus* collected from the wild (Wheeler 1967; More 1970; Parker and Holcomb 1973; Elam and Green 1974). However, the labor-intensive process of collecting wild postlarvae was abandoned by the early 1970's when hatchery-reared postlarvae became available.

Within a few years, Texas researchers had identified many of the fundamental variables of shrimp pond management, such as: (1) importance of predator and competitor control (Parker et al. 1972); (2) usefulness of nursery ponds (Elam and Green 1974); (3) value of pond fertilization (Wheeler 1968; Rubright et al. 1981); (4) need for nutritious and adequately bound feeds (Parker and Holcomb 1973; Elam and Green 1974); (5) avoidance of oxygen depletion through water exchange and aeration (Wheeler 1968); (6) mortality associated with black anaerobic pond bottoms (Wheeler 1968; More 1970); (7) efficiency of harvesting by draining rather than seining (Holcomb and Parker 1973); (8) higher tolerance of *P. duorarum* than of *P. aztecus* and *P. setiferus* to winter temperatures (Parker and Holcomb 1973; Hysmith and Elam 1974); and (9) potential for polyculture of shrimp with oysters (More 1970) or fish (Rossberg and Strawn 1980).

Survival, growth, and yield during pond grow-out were found to vary with the species of shrimp utilized. Of the three native species, *P. setiferus* outperformed *P. aztecus* (Parker and Holcomb 1973; Elam and Green 1974), and *P. aztecus* out-performed *P. duorarum* (More 1970). However, more significant was the 1972 finding that a non-native species, *P. vannamei*, produced higher yields than either *P. setiferus* or *P. aztecus*. Subsequent trials with high densities of *P. vannamei* in aerated experimental ponds yielded record harvests of approximately 5 MT/ha (Parker et al. 1974). *Penaeus stylirostris*, another non-native species, was later shown to perform well in polyculture with *P. vannamei* (Chamberlain et al. 1981; Huang 1983).

Although pond grow-out trials with *P. vannamei* and *P. stylirostris* were encouraging, the difficulty of routinely importing postlarvae from their native habitat on the Pacific coast of Latin America became a major stumbling block to commercial development. Larval rearing techniques previously developed at the NMFS Galveston Laboratory relied upon offshore collection of fully mature females. Techniques had not been developed for inducing pond-raised shrimp to mature and spawn in captivity. In order to take advantage of the superior performance of non-native species, it became imperative to shift research emphasis back to the hatchery to study techniques for controlling reproduction of shrimp in captivity.

Reproduction in Captivity

It was not until the late 1970's that facilities were built to induce ovarian maturation of shrimp in captivity. Principal participants in this research were the NMFS Galveston Laboratory (Clark and Lynn 1977; Brown et al. 1979; Brown et al. 1980; Brown et al. 1984), Texas A&M University System (Conte et al. 1978; Chamberlain and Lawrence 1981a, b; Leung-Trujillo and Lawrence 1985), and the University of Houston (Middleditch et al. 1979 1980). By the early 1980's, adequate techniques were developed for maturing and spawning a variety of penaeid species. This technology stimulated development of commercial hatcheries and provided an opportunity for genetic selection studies to begin (Lester 1983, 1988).

Extending the Growing Season

The temperate growing season in Texas prevents year-round culture of shrimp in outdoor ponds. Researchers have attempted to alleviate this competitive disadvantage by raising cold-tolerant species, such as *P. duorarum* (Hysmith and Elam 1974) or *P. japonicus* (unpublished data 1980) through the winter or by using thermal effluent from power stations (Gould et al. 1973; Chamberlain et al. 1980; Huang 1983). Other studies have cultured shrimp at high density in indoor raceways either as a means of headstarting the growing season during the early spring or as an independent system for year-round indoor grow-out (Mock et al. 1973, 1978; Sturmer and Lawrence 1986, 1987; Arnold et al. 1987).

Supporting Research

Many of the advancements described above could not have been accomplished without accompanying research on diseases, nutrition, physiology, seafood technology, and economics. Examples of disease research include identifications by the NMFS Galveston Laboratory and Texas A&M University of: bacterial and fungal infestations in larval shrimp (Lightner and Fontaine 1973; Lightner and Lewis 1975), pathogens responsible for gill disease in intensive culture systems (Lightner et al. 1975), and major parasites and diseases in wild (Fontaine 1985) and cultured shrimp (Lightner 1975; Johnson 1978). Pathologists also have developed techniques for managing and controlling diseases (Johnson 1974; Lewis and Lawrence 1983).

Shrimp nutrition research in Texas began with the broad work of Shewbart et al. (1973) and Zein-Eldin and Meyers (1973). Subsequent laboratory studies investigated requirements for and digestibility of protein, lipid, and carbohydrate (Fenucci et al. 1980, 1982; Smith et al. 1985; Lee and Lawrence 1985; Akiyama et al. 1989). Pond studies evaluated the utilization of prepared versus natural feeds (Rubright et al. 1981; Anderson et al. 1987).

Physiological research has played an important role in determining environmental tolerances to temperature, salinity, and dissolved oxygen (Zein-Eldin 1963; Zein-Eldin and Griffith 1969; Keiser and Aldrich 1976; Seidman and Lawrence 1985). It has also elucidated basic processes such as egg fertilization (Clark et al. 1973; Clark and Lynn 1977), and hormonal control of shrimp reproduction and molting (Quackenbush and Keeley 1986; Chan et al. 1988; Rankin et al. 1989).

Seafood technology research in Texas has provided information about mechanisms responsible for shrimp spoilage, flavor changes, black spot, and other factors which affect the quality and value of shrimp (Cobb et al. 1974; Vanderzant et al. 1974; Bottino et al. 1979; McCoid et al. 1984; Finne and Miget 1985). This information has been helpful in developing suitable methods for processing and storing highly marketable pond-raised shrimp.

Economic analyses of the potential for shrimp farming in Texas have been published periodically to evaluate changes in production technology (Adams et al. 1980; Griffin et al. 1981; Hanson et al. 1985: Juan et al. 1988).

History of Commercial Shrimp Farming in Texas

The success of early larval rearing and pond grow-out research during the late 1960's encouraged Dow Chemical to initiate commercial hatchery, grow-out, and nutrition work during the early 1970's. Harry Cook and Harvey Persyn operated the Dow Chemical hatchery near Freeport, Texas. Unfortunately, that venture proved to be premature.

The inability to control reproduction of penaeid shrimp during the 1970's led several U.S. groups to investigate culture of freshwater shrimp (*Macrobrachium rosenbergii*), which spawns readily in captivity. Sun Oil Company under the guidance of Dr. Arch Stuart and Mr. Durwood Dugger started a subsidiary called Aquaprawns in Port Isabel, Texas, in 1976 to begin pilot trials with *M. rosenbergii*. That project was bought by Commercial Shrimp Culture International (CSCI) in 1978 and work was begun on intensive culture of *P. vannamei* in 15-m diameter round tanks with artificial substrate and center drains. Hurricane Allen demolished the CSCI facility in August 1980. After denial of permits for development of a marine site on the Brownsville ship channel, CSCI rebuilt their facility at an inland location near Los Fresnos, Texas, the following year using federal disaster relief funds. At the new 65-acre pond site, CSCI primarily undertook *Macrobrachium* culture, which proved to be impractical, largely because of marketing problems. During 1983 and 1984, a major portion of CSCI's low-salinity pond system was devoted to penaeid culture. Unfortunately, postlarval availability at that time was poor. Consequently, ponds were stocked with *P. setiferus* and *P. vannamei* late in the growing season and at low density, resulting in production of only 600-900 kg/ha. CSCI was purchased by another group in 1985 and eventually used for culture of red drum.

Dr. Jack Parker resigned from Texas A&M University in 1980 and founded Laguna Madre Shrimp Farms near Bayview, Texas (Fig. 1). This farm began in 1981 with a 3-phase, 28-ha, semi-intensive pond system and a pilot-scale greenhouse-covered hatchery. A second 28-ha pond module was added in 1982. The pilot-scale greenhouse-covered hatchery, managed by Dr. Ben Ribelin, began producing sufficient postlarvae for stocking all the grow-out ponds at Laguna Madre Shrimp Farms in 1982. This was a

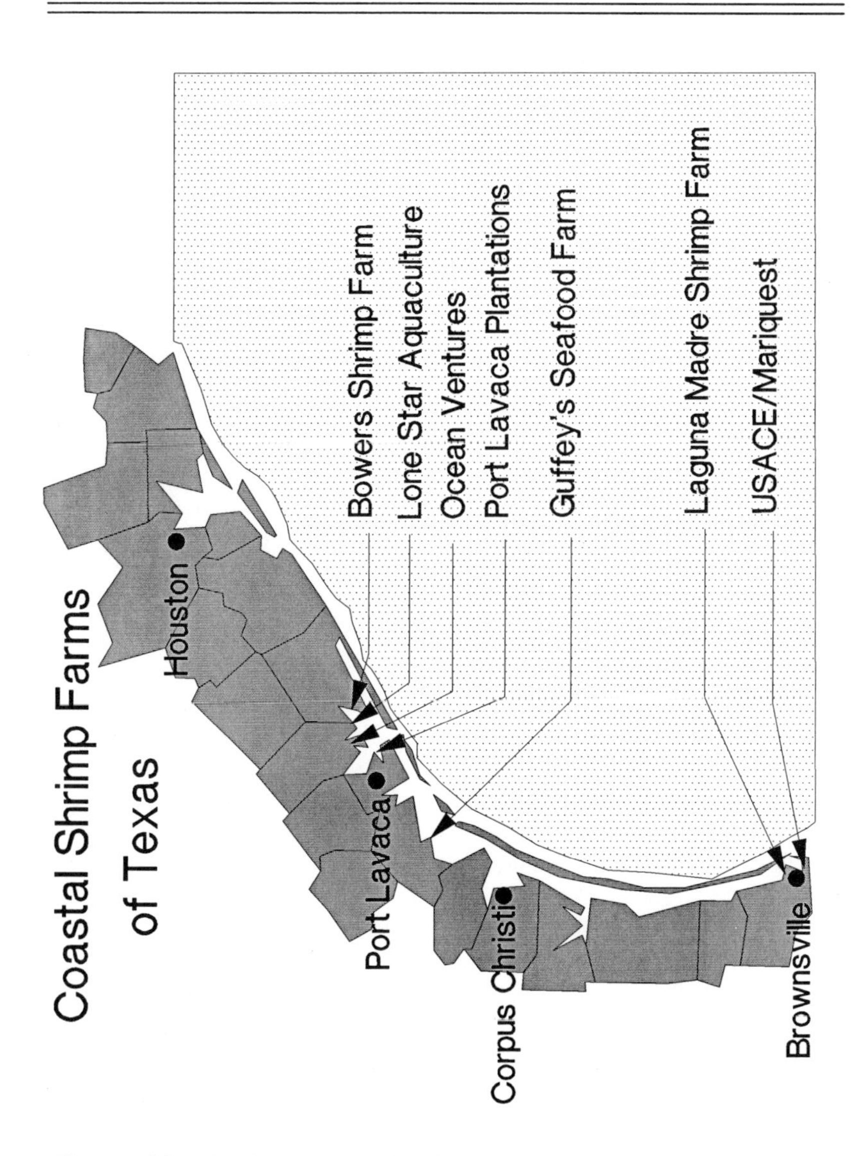

Figure 1. Map showing coastal shrimp farms in Texas.

major turning point in the history of Texas shrimp farming, because it marked the first time that postlarval availability did not limit pond stocking. In 1983, Laguna Madre Shrimp Farms expanded its pond area to 182-ha of water surface and constructed a large permanent hatchery. Production from Laguna Madre's ponds, which average about 8-ha in size, has increased from around 1900 kg/ha/yr to about 2900 kg/ha/yr through the use of aeration and more intensive management.

Several out-of-state hatcheries began selling postlarvae to Texas growers in the mid 1980's. The commercial availability of postlarvae greatly diminished the capital cost and risk of entering the shrimp farming business. As a result, one group attempted during 1984 to raise hatchery-reared *P. setiferus* to market size in a 62-ha impoundment near Anahuac, Texas. A different group utilized the same facility for culture of *P. vannamei* in 1985. This low-input system fared poorly for both groups, primarily due to difficulty in harvesting the shrimp. Several groups responded to the availability of postlarvae by building relatively small intensive pond systems. These included Ocean Ventures (1985), Guffey Seafood Farms (1987), Port Lavaca Shrimp Farms (1988), and Bowers Shrimp Farm (1989). These groups have attempted with varying degrees of success to develop intensive management techniques appropriate to the costs of land, labor, climate, and energy in Texas. Although intensive farms are known to be profitable in areas such as Taiwan and the Philippines, most growers feel that direct transplantation of intensive culture techniques from such areas would not be profitable in Texas. Instead, Texas growers have adopted a combination of diverse techniques such as fertilization procedures from striped bass fingerling growers, feeding tray technology from Taiwan, paddlewheel aeration and feed distribution methods from the catfish farming industry, and fish-pump harvesting methods used in the trout industry. Results have been improving, but no shrimp farm in Texas has been more than marginally profitable as of 1989. Of the grow-out systems attempted in Texas thus far, intensive ponds seem to show the greatest potential. Projections indicate that the total area of intensive shrimp ponds in Texas is steadily increasing (Fig. 2). Most Texas farms have been converting an increasing proportion of their ponds from semi-intensive to intensive management for the last several years. Each of these farms is anticipating intensive yields of 3-4 MT/ha during 1989.

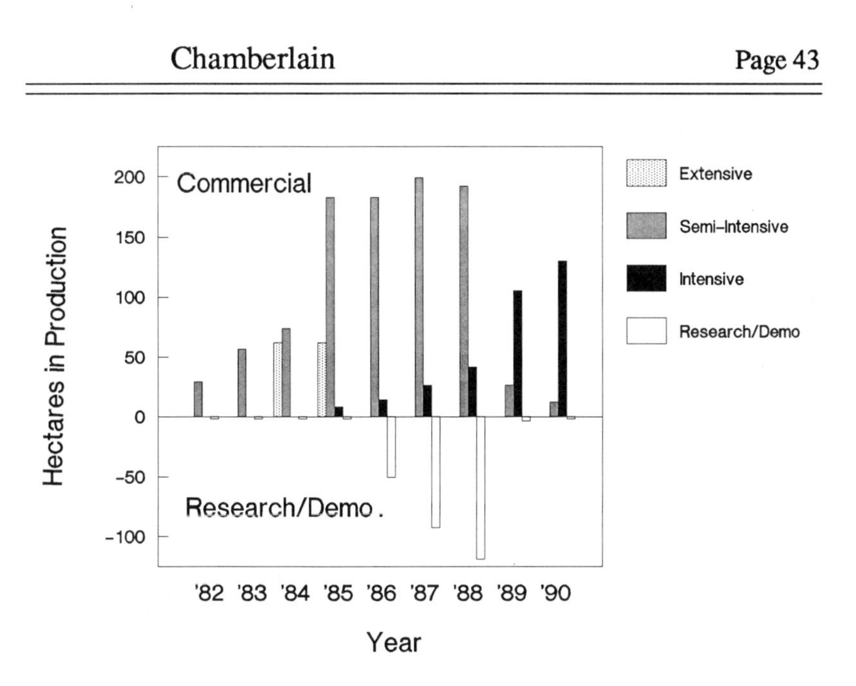

Figure 2. Graph of total area devoted to shrimp ponds in Texas.

In addition to the coastal farms described above, several groups in Texas began experimenting with use of underground brine water for culture of shrimp at inland sites. For example, Ben Kirsch in collaboration with Texas A&M University demonstrated production of *P. vannamei* and *P. monodon* at an inland location near Raymondville, Texas (Sal Del Rey Ranch). Several small farms also have been operating in West Texas using underground brine (e.g., Genesis Seafood, Stanton, Texas).

The newest culture system entering the commercial shrimp farming arena in Texas is the recirculating, indoor tank system. Dr. Bob Brick managed an early attempt at a commercial indoor system, King James Shrimp Company, in the Chicago area during the early 1980's. He and Dr. Bud Geiger have recently announced plans for their company, Texas Mariculture, Inc. (TMI), to develop a commercial prototype indoor system in Bryan, Texas, based on earlier research of TMI at San Marcos, Texas.

CURRENT STATUS OF COMMERCIAL SHRIMP CULTURE

Principal Species

Virtually all of the shrimp commercially cultured in Texas during 1988 were *P. vannamei*. Commercial production with native species has yet to produce satisfactory yields. Consequently, these species are no longer utilized. However, there is a possibility that these species may regain attention in the culture of bait shrimp for the recreational fishing industry. Several producers have previously polycultured *P. stylirostris* with *P. vannamei* as described by Chamberlain et al. (1981). Also, small-scale trials with *P. monodon* and *P. penicillatus* were conducted during 1988.

Area in Production

The pond area devoted to commercial shrimp culture has fluctuated since 1982 from a low of 29 ha to a high of 234 ha (Fig. 2). Over 100 ha of ponds are expected to lay idle in 1989 while farms conduct various trials. Projections for 1990 indicate that somewhat fewer ponds will be fallowed.

Farms in Planning Stage

At least two farms are now in the planning stage. One of these (182 ha of water surface) is now being constructed on the Arroyo Colorado River near its confluence with the Lower Laguna Madre.

Total Production

Total production of shrimp fluctuated around 500 MT during 1987 and 1988 (Fig. 3). Production declined markedly during 1989 as a reflection of fallowed pond area and the termination of two large public sector research/demonstation projects. Production during 1990 is projected to return to approximately 500 MT as existing ponds are operated more intensively and a new farm comes on line.

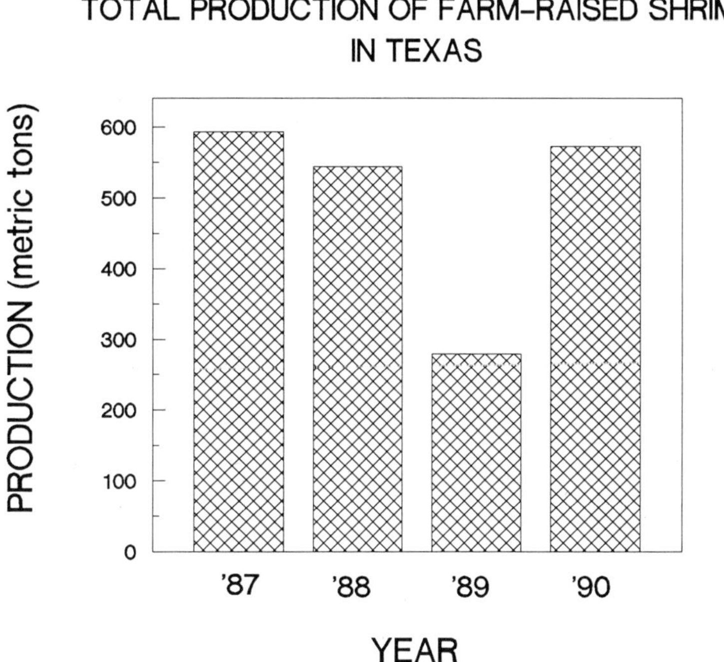

Figure 3. Graph of total farm-raised shrimp production in Texas.

Number of Crops per Year

With the exception of Ocean Ventures and USACE/Mariquest, shrimp farms in Texas attempt only one crop per year. Ocean Ventures has installed greenhouse structures over portions of their 2.2-acre grow-out ponds to allow early stocking of postlarvae and harvest of two crops during the 6-7 month warm season. USACE/Mariquest has produced two crops per year through the use of an uncovered nursery pond. Prospects of producing two crops of small shrimp rather than one of larger shrimp during the warm season are attractive because small shrimp have a lower food conversion than larger shrimp. Also, production of two crops per year would improve cash flow and reduce the risk of crop loss due to disease or hurricane.

However, these advantages must be weighed against the double cost of postlarvae and the relatively low value of small shrimp with two crops per year. Efforts to produce two crops per year are expected to increase as improvements in technology promote more rapid growth in ponds.

USACE/Mariquest has also experimented with a second crop of cold-tolerant shrimp (*P. penicillatus*) during the winter, but results were unsatisfactory, partly because of a record freeze.

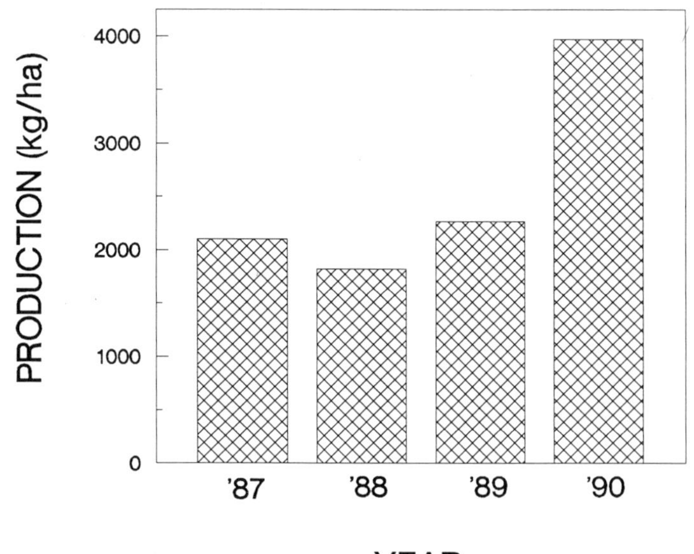

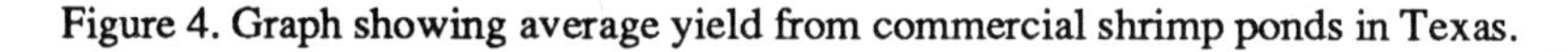
Figure 4. Graph showing average yield from commercial shrimp ponds in Texas.

Average Production

Average production of commercial ponds diminished somewhat during 1988 due to a variety of both explained and unexplained problems (Fig. 4).

Some of these were apparently related to inferior postlarval quality and to problems encountered by excessively increasing the stocking density of semi-intensive ponds. Production trials during 1989 showed a slight increase in average yield due to greater use of intensive management. During 1990, production is expected to increase considerably as nearly all ponds will be equipped with aeration.

Principal Size Grades of Product and Target Markets

Most shrimp harvested from Texas ponds fall in the 31-40 count (tails per pound) category, although both larger and smaller shrimp are produced. Most smaller farms sell their head-on production directly to a coastal processor for processing and marketing. However, a few farms contract with processing plants only to have their product processed and packaged. Then, the farms manage marketing of their custom-packaged shrimp. Several shrimp farms are evaluating alternate marketing approaches to avoid direct competition with commodity-priced imports.

Number, Status, and Production of Hatcheries

As of 1989, Texas had only one commercial shrimp hatchery, Laguna Madre Shrimp Farms. This hatchery has a production capability of 25 million postlarvae per month. It is capable of operating without use of wild broodstock by relying on pond-raised animals.

Several farms have expressed an interest in beginning new hatcheries in the near future. Their primary interest is to guarantee quality and supply of seed and secondarily to reduce the price.

Dollar Value of Industry

For the last 2 years the farm-gate value of the Texas shrimp farming industry has fluctuated around $ 3 million. It is expected to hold this value during 1990 and to exceed $5.0 million in 1991 when a new farm comes on line and fallowed ponds are put back into production.

Research and Extension Support

Several universities including Texas A&M University, University of Texas, and University of Houston are actively engaged in shrimp aquaculture research. Much of the funding for this effort has historically originated from the Texas A&M University Sea Grant Program. Recently, additional funds have been allocated for shrimp farming research in Texas by the U.S. Department of Agriculture through an interstate consortium of researchers.

The Texas Agricultural Extension Service (TAEX), a branch of the Texas A & M University System, supports commercial shrimp farming development through short courses, manuals (Johnson 1978; Chamberlain et al. 1985), the Coastal Aquaculture newsletter, start-up assistance, disease diagnosis, and so forth. Marine Advisory Service Agents located in coastal counties of Texas provide personal assistance in locating suitable property and interfacing with aquaculture, engineering, disease, and marketing specialists within TAEX.

The General Land Office of Texas has recently assisted aquaculture development by leasing state-owned property to a shrimp producer in the Matagorda Bay area.

Major Problems

The trend toward ever-increasing densities of shrimp in ponds has led to reductions in growth and survival at several farms. It is critical that farms implement cost effective means of aeration and water circulation to control dissolved oxygen and fouling of the pond bottom. This represents a major capital expense.

Serious questions have been raised about the effect of shrimp viruses on growth and survival. This is a controversial area, because the techniques for isolating and identifying some viruses are imperfect at this time.

Prospects for the Future

The recent trends in growth of intensive farms in Texas are encouraging. If increases in yields can be achieved as projected by commercial growers and if high-value niche markets can be developed, then the industry is expected to expand. Total area available for coastal pond construction in Texas is at least 12,000 ha. Inland property is not expected to be widely developed until the optimal coastal sites are taken.

LITERATURE CITED

Adams, C. M., W. L. Griffin, J. R. Nichols, and R. W. Brick. 1980. Bioengineering-economic model for shrimp mariculture systems 1979. Publication TAMU-SG-203, Texas A&M University Sea Grant College Program, College Station, Texas, 118 p.

Akiyama, D. M., S. R. Coelho, A.L. Lawrence, and E. H. Robinson. 1989. Apparent digestibility of feedstuffs by the marine shrimp, *Penaeus vannamei* Boone. Nippon Suisan Gakkaishi (formerly Bulletin of the Japanese Society of Scientific Fisheries) 55:91-98.

Anderson, R. K., A. L. Lawrence, and P. L. Parker. 1987. A 13C/12C tracer study of the utilization of presented feed by a commercially important shrimp, *Penaeus vannamei*, in a pond grow-out system. Journal of the World Aquaculture Society 18: 148-155.

Arnold, C. R., B. Brawner, and M. Pattillo. 1987. High density recirculating grow-out systems. Pages V-53 to V-58 *in* G.W. Chamberlain, R.J. Miget, and M. G. Haby, editors. Manual on red drum aquaculture, Texas Agricultural Extension Service Publication, Corpus Christi, Texas.

Bottino, N. R., M. L. Lilly, and G. Finne. 1979. Fatty acid stability of Gulf of Mexico brown shrimp (*Penaeus aztecus*) held on ice and in frozen storage. Journal of Food Science 44(6):1778-1779.

Brown, A., Jr., J. McVey, B. S. Middleditch, and A. L. Lawrence. 1979. Maturation of white shrimp (*Penaeus setiferus*) in captivity. Proceedings of the World Mariculture Society 10:435-444.

Brown, A., Jr., J. P. McVey, B. M. Scott, T. D. Williams, B. S. Middleditch, and A. L. Lawrence. 1980. The maturation and spawning of *Penaeus stylirostris* under controlled conditions. Proceedings of the World Mariculture Society 11:488-499.

Brown, A., Jr., D. Tave, T. D. Williams, and M. J. Duronslet. 1984. Production of second generation penaeid shrimp, *Penaeus stylirostris*, from Mexico. Aquaculture 41: 81-84.

Chamberlain, G. W., D. L. Hutchins, A. L. Lawrence, and J. C. Parker. 1980. Winter culture of *Penaeus vannamei* in ponds receiving thermal effluent at different rates. Proceedings of the World Mariculture Society 11:30-43.

Chamberlain, G. W., D. L. Hutchins, and A. L. Lawrence. 1981. Mono- and polyculture of *Penaeus vannamei* and *Penaeus stylirostris*. Proceedings of the World Mariculture Society 12(1):209-224.

Chamberlain, G. W. and A. L. Lawrence. 1981a. Maturation, reproduction, and growth of *Penaeus vannamei* and *P. stylirostris* fed natural diets. Journal of the World Mariculture Society 12(1):209-224.

Chamberlain, G. W. and A. L. Lawrence. 1981b. Effect of light intensity and male and female eyestalk ablation on reproduction of *Penaeus stylirostris* and *P. vannamei*. Journal of the World Mariculture Society 12(2):357-372.

Chamberlain, G.W., M.G. Haby, and R.J. Miget, editors. 1985. Texas shrimp farming manual: an update on current technology. Texas Agricultural Extension Service, Texas A & M University System, Corpus Christi, TX, 273pp.

Chan, S-M, S.M. Rankin, and L. L. Keeley. 1988. Characterization of the molt stages in *Penaeus vannamei*: setogenesis and hemolymph levels of total protein, ecdysteroids, and glucose. Biological Bulletin 175:185-192.

Clark, W. H. and J. W. Lynn. 1977. A Mg^{++} dependent cortical rod reaction in the eggs of penaeid shrimp. The Journal of Experimental Zoology 200(1):177-183.

Clark, W. H., Jr., P. Talbot, R. A. Neal, C. R. Mock, and B. R. Salser. 1973. In vitro fertilization with non-motile spermatozoa of the brown shrimp *Penaeus aztecus*. Marine Biology 22:353-354.

Cobb, B. F., C. Vanderzant, and K. Hyder. 1974. Effect of ice storage upon the free amino acid contents of tails of white shrimp (*Penaeus setiferus*). Agricultural and Food Chemistry 22(6):1052-1055.

Conte, F. S., M. J. Duronslet, W. H. Clark, and J. C. Parker. 1978. Maturation of *Penaeus stylirostris* (Stimpson) and *P. setiferus* (Linn.) in hypersaline water near Corpus Christi, Texas. Proceedings of the World Mariculture Society 8:327-334.

Cook, H. L. 1969. A method of rearing penaeid shrimp larvae for experimental studies. FAO Fisheries Report 57 (3):709-715.

Cook, H. L. and M. A. Murphy. 1966. Rearing penaeid shrimp from eggs to postlarvae. Proceedings of the 19th Annual Conference of the Southeastern Association of Game and Fish Commissioners 19:283-288.

Elam, L. L. and A. W. Green. 1974. Culture of white shrimp (*Penaeus setiferus* Linnaeus) in static water ponds. Proceedings of the World Mariculture Society 5:87-97.

Fenucci, J., Z. Zein-Eldin, and A. L. Lawrence. 1980. The nutritional response of two penaeid species to various levels of squid meal in a prepared feed. Proceedings of the World Mariculture Society 11:403-409.

Fenucci, J. L., A. C. Fenucci, A. L. Lawrence, and Z. P. Zein-Eldin. 1982. The assimilation of protein and carbohydrate from prepared diets by the shrimp, *Penaeus stylirostris*. Journal of the World Mariculture Society 13:134-145.

Finne, G. and R. Miget. 1985. The use of sulfiting agents in shrimp - US regulations for domestically produced and imported products. Infofish Marketing digest 4:39-40.

Fontaine, C. T. 1985. A survey of potential disease-causing organisms in bait shrimp from West Galveston Bay, Texas. NOAA Technical Memorandum NMFS-SEFC-169, 41 pp.

Griffin, W.L. , J. S. Hanson, R. W. Brick, and M. A. Johns. 1981. Bioeconomic modeling with stochastic elements in shrimp culture. Journal of the World Mariculture Society 12(1):94-103.

Gould, R. A., D. V. Aldrich, and C. R. Mock. 1973. Experimental pond culture of brown shrimp (*Penaeus aztecus*) in power plant effluent water. Proceedings of the World Mariculture Society 4:195-213.

Hanson, J. S., W.L. Griffin, J.W. Richardson, and C.J. Nixon. 1985. Economic feasibility of shrimp farming in Texas: an investment analysis for semi-intensive grow-out. Journal of the World Mariculture Society 16:129-150.

Holcomb, H. W., Jr., and J. C. Parker. 1973. Efficiency of drain and seine harvest techniques in experimental penaeid shrimp culture ponds. Proceedings of the World Mariculture Society 4:215-234.

Huang, H. 1983. Factors affecting the successful culture of *Penaeus stylirostris* and *Penaeus vannamei* at an estuarine power plant site: temperature, salinity, inherent growth variability, damselfly nymph predation, population density and distribution, and polyculture. Ph.D. dissertation, Texas A&M University, College Station, Texas. 221 p.

Hudinaga, M. 1942. Reproduction, development, and rearing of *Penaeus japonicus* Bate. Japanese Journal of Zoology 10:305-393.

Hysmith, B. T. and L. L. Elam. 1974. Job 1: Penaeid shrimp culture. Abstract of Completion Report, Texas Parks and Wildlife Department. 5 pp.

Johnson, S. K. 1974. Toxicity of several management chemicals to penaeid shrimp. Texas Agricultural Extension Service Fish Disease Diagnostic Laboratory Publication Number FDDL-13, 10 pp.

Johnson, S. K. 1978. Handbook of shrimp diseases. Texas A&M University Sea Grant Publication Number TAMU-SG-75-603. 23 pp.

Juan, Y.-S., W.L. Griffin, and A.L. Lawrence. 1988. Production costs of juvenile penaeid shrimp in an intensive greenhouse raceway nursery system. Journal of the World Aquaculture Society 19:149-160.

Keiser, R. K., Jr. and D. V. Aldrich. 1976. Salinity preference of postlarval brown and white shrimp (*Penaeus aztecus* and *P. setiferus*) in gradient tanks. Texas A&M University Sea Grant Publication TAMU-SG-75-208. 260 pp.

Kuban, F.D., A.L. Lawrence, and J. S. Wilkenfeld. 1985. Survival, metamorphosis, and growth of four larval penaeid species fed six food combinations. Aquaculture 47:151-162.

Lee, P. G. and A. L. Lawrence. 1985. Effects of diet and size on growth, feed digestibility, and digestive enzyme activities of the marine shrimp, *Penaeus setiferus*. Journal of the World Mariculture Society 16: 275-287.

Lester, L. J. 1983. Developing a selective breeding program for penaeid shrimp mariculture. Aquaculture 33:41-50.

Lester, L. J. 1988. Differences in larval growth among families of *Penaeus stylirostris* Stimpson and *P. vannamei* Boone. Aquaculture and Fisheries Management 19:243-251.

Leung-Trujillo, J.R. and A.L. Lawrence. 1985. The effect of eyestalk ablation on spermatophore and sperm quality in *Penaeus vannamei*. Journal of the World Mariculture Society 16:258-266.

Lewis, D. H. and A. L. Lawrence. 1983. Immunoprophylaxis to *Vibrio* sp. in pond reared shrimp. Proceedings of the International Conference on Warm Water Aquaculture Crustacea 1:304-307.

Lightner, D. V. 1975. Some potentially serious disease problems in the culture of penaeid shrimp in North America. Pages 75-97 *in* Proceedings of the 3rd U. S. - Japan Meeting on Aquaculture, Tokyo, Japan, October 15-16 1974.

Lightner, D. V. and C. T. Fontaine. 1973. A new fungus of the white shrimp *Penaeus setiferus*. Journal of Invertebrate Pathology 22:94-99.

Lightner, D. V. and D. H. Lewis. 1975. A septicemic bacterial disease syndrome of penaeid shrimp. Marine Fisheries Review 37:25-28.

Lightner, D. V., C. T. Fontaine, and K. Hanks. 1975. Some forms of gill disease in penaeid shrimp. Proceedings of the World Mariculture Society 6:347-365.

McCoid, V., R. Miget, and G. Finne. 1984. Effect of environmental salinity on the free amino acid composition and concentration in penaeid shrimp. Journal of Food Science 49:327-330.

Middleditch, B. S., S. R. Missler, H. B. Hines, E. S. Chang, J. P. McVey, A. Brown, and A. L. Lawrence. 1980. Maturation of penaeid shrimp: lipids in the marine food web. Proceedings of the World Mariculture Society 11:463-470.

Middleditch, B. S., S. R. Missler, D. G. Ward, J. B. McVey, A. Brown, and A. L. Lawrence. 1979. Maturation of penaeid shrimp: dietary fatty acids. Proceedings of the World Mariculture Society 10:472-476.

Mock, C. R. and M. A. Murphy. 1970. Techniques for raising penaeid shrimp from the egg to postlarvae. Proceedings of the World Mariculture Society 1: 143-156.

Mock, C. R., R. A. Neal, and B. R. Salser. 1973. A closed raceway for the culture of shrimp. Proceedings of the World Mariculture Society 4:247-259.

Mock, C. R., D. B. Revera, and C. T. Fontaine. 1980. The larval culture of *Penaeus stylirostris* using modifications of the Galveston Laboratory technique. Proceedings of the World Mariculture Society 11:102-117.

Mock, C. R., L. R. Ross, and B. R. Salser. 1978. Design and preliminary evaluation of a closed system for shrimp culture. Proceedings of the World Mariculture Society 8:335-369.

More, W. R. 1970. Saltwater pond progress report No. 1, December 1968-November 1969. Coastal Fisheries, Texas Parks and Wildlife Department, Austin, Texas. 20 p.

Parker, J. C., F. S. Conte, W. S. McGrath, and B. W. Miller. 1974. An intensive culture system for penaeid shrimp. Proceedings of the World Mariculture Society 5:65-79.

Parker, J. C. and H. W. Holcomb, Jr. 1973. Growth and production of brown and white shrimp (*Penaeus aztecus* and *P. setiferus*) from experimental ponds in Brazoria and Orange counties, Texas. Proceedings of the World Mariculture Society 4:215-234.

Parker, J. C., H. W. Holcomb, Jr., W. G. Klussman, and J. C. McNeill IV. 1972. Effect of fish removal on the growth and condition of white shrimp in brackish ponds. Proceedings of the World Mariculture Society 3:287-297.

Quackenbush, L. S. and L. L. Keeley. 1986. Vitellogenesis in the shrimp, *Penaeus vannamei*. American Zoologist 26: 56A.

Rankin, S. M., J. Y. Bradfield, and L. L. Keeley. 1989. Ovarian protein synthesis in the South American white shrimp, *Penaeus vannamei*, during the reproductive cycle. International Journal of Invertebrate Reproduction and Development 15:27-33.

Rossberg, K. S. and R. K. Strawn. 1980. Comparative growth and survival of brown shrimp cultured with Florida pompano, black drum, and striped mullet. Proceedings of the World Mariculture Society 11:219-225.

Rubright, J. S., J. L. Harrell, H. W. Holcomb, and J. C. Parker.1981. Responses of planktonic and benthic communities to fertilizer and feed applications in shrimp mariculture. Proceedings of the World Mariculture Society 12(1):281-299.

Seidman, E. and A. L. Lawrence. 1985. Growth, feed digestibility and proximate body composition of juvenile *Penaeus vannamei* and *Penaeus monodon* grown at different dissolved oxygen levels. Journal of the World Mariculture Society 16: 275-287.

Shewbart, K. L., W. L. Mier, and P. D. Ludwig. 1973. Nutritional requirements of brown shrimp, *Penaeus aztecus*. Texas A&M University Sea Grant Publication, Number TAMU-SG-73-205, 53 pp.

Smith, L. L., P. G. Lee, A. L. Lawrence, and K. Strawn. 1985. Growth and digestibility by three sizes of *Penaeus vannamei* Boone: effect of dietary protein level and protein source. Aquaculture 46:85-96.

Sturmer, L. N. and A. L. Lawrence. 1986. Evaluation of raceways as intensive nursery rearing systems for penaeid shrimp. Abstract, Seventeenth Annual Meeting of the World Aquaculture Society, Reno, Nevada.

Sturmer, L. N. and A. L. Lawrence. 1987. Effects of stocking density on growth and survival of *P. vannamei* and *P. stylirostris* postlarvae in intensive nursery raceways. Abstract, Eighteenth Annual Meeting of the World Aquaculture Society, Guayaquil, Ecuador.

Vanderzant, C., B. F. Cobb, and R. Nickelson. 1974. Role of microorganisms in shrimp quality: a research summary. Texas A&M University Sea Grant Publication Number TAMU-SG-74-201. 46 pp.

Wheeler, R.S. 1967. Experimental rearing of postlarval brown shrimp to marketable size in ponds. Commercial Fisheries Review 29 (3): 49-52.

Wheeler, R.S. 1968. Culture of penaeid shrimp in brackish-water ponds 1966-67. Proceedings of the 22nd Annual Conference of the Southeastern Association of Game and Fish Commissioners 22:387-391.

Wilkenfeld, J. S., A.L. Lawrence, and F.D. Kuban. 1983. Rearing penaeid shrimp larvae in a small-scale system for experimental purposes. Pages 72-81 *in* G.L. Rogers, R. Day, and A. Lim, editors. Proceedings of the 1st International Conference on Warm Water Aquaculture, Crustacea. Feb. 9-11 1983, Brigham Young University, Laie, Hawaii.

Zein-Eldin, Z. P. 1963. Effect of salinity on the growth of post-larval penaeid shrimp. Biological Bulletin (Woods Hole) 125(1):188-196.

Zein-Eldin, Z. P. and S. P. Meyers. 1973. General considerations of problems in shrimp nutrition. Proceedings of the World Mariculture Society 4:299-317.

Zein-Eldin, Z. P. and G. W. Griffith. 1969. An appraisal of the effects of salinity and temperature on growth and survival of postlarval penaeids. FAO Fishery Report 57(3):1015-1026.

SHRIMP CULTURE IN NORTH AMERICA AND THE CARIBBEAN: HAWAII 1988

Gary D. Pruder

ABSTRACT

Shrimp culture activities based in Hawaii include research, production, and international consulting. During 1988, research increased, production decreased, and international consulting remained steady. Total revenues generated by these activities were approximately $7.5 million in 1988.

Shrimp research at The Oceanic Institute and the University of Hawaii expanded through increased support from the CSRS/USDA for the U.S. Marine Shrimp Farming Program and the Center for Tropical and Subtropical Aquaculture. Level funded shrimp research support to the University was provided by the Office of Sea Grant and the Aquaculture Development Program of the State of Hawaii. Most research activities were dedicated to reduced risk and increased profitability for marine shrimp farming in the United States.

Marine Culture Enterprises, one of two major marine shrimp producers in Hawaii, ceased operation in 1988. Amorient Aquaculture International, the other major producer, began expansion of its operations. There are five new marine shrimp business start-ups which are attempting marine shrimp production: Aurea Marine, Kahuku Shrimp Company, Molokai Seafarms, Ohia Shrimp Farm, and Pacific Sea Farms.

Aquatic Farms continues as a major marine shrimp international consultant. Aquacultural Concepts and the Hawaii Aquaculture Company have international consulting contracts in marine shrimp farming. These and several other consulting companies are involved in feasibility studies for additional marine shrimp farms in Hawaii. Hawaii continues to experience promising expansions and disappointing setbacks in marine shrimp culture. However, never before have so many national resources and collective expertise been coordinated and committed to expand domestic marine shrimp farming. Cooperation between research institutions, commercial farmers, and Federal and State agencies is excellent. Advances in production systems, maturation and reproduction, disease control, processing, marketing, and economics strengthen our resolve to develop a competitive domestic marine shrimp farming industry.

INTRODUCTION

Aquaculture in general and especially marine shrimp farming fit into the State of Hawaii's overall strategy for economic development and diversification. It has the potential to use underutilized resources, e.g., brackish water, saltwater, agricultural wastes, and lava lands. The Governor's Aquaculture Industry Development Committee (Corbin 1984) reported that industry expansion was dependent upon reducing or eliminating constraints identified in the areas of management, technology and knowledge, marketing and economics, water and land, and government activities (development climate). In comparison, a 1984 poll of private Hawaiian aquaculture companies (Pruder et al. 1985) indicated that the principal constraints were government regulations, the cost of financing, and the availability and cost of postlarvae.

Research efforts in 1988, as in 1986 and 1987, were directed specifically to increase rates of return, decrease risks, provide the necessary seed, and minimize the impact of regulations concerning the use of chemotherapeutics and the discharge of aquacultural effluent. Results from such research are only now beginning to impact the private sector. Therefore, the factors that influenced the commercial marine shrimp farms in 1988 were in place and relatively unchanged from the Governor's Study in 1984. Consulting companies in general are exporting technologies that would probably not be cost effective or competitive in Hawaii.

The future of shrimp culture in Hawaii depends upon the persistence of entrepreneurs; the continued development, transfer, and utilization of intensive nursery and growout technology; adoption of disease control procedures, including reliable sources of specific-pathogen-free shrimp broodstock and postlarvae; and reduction of chemotherapeutic and effluent discharge constraints.

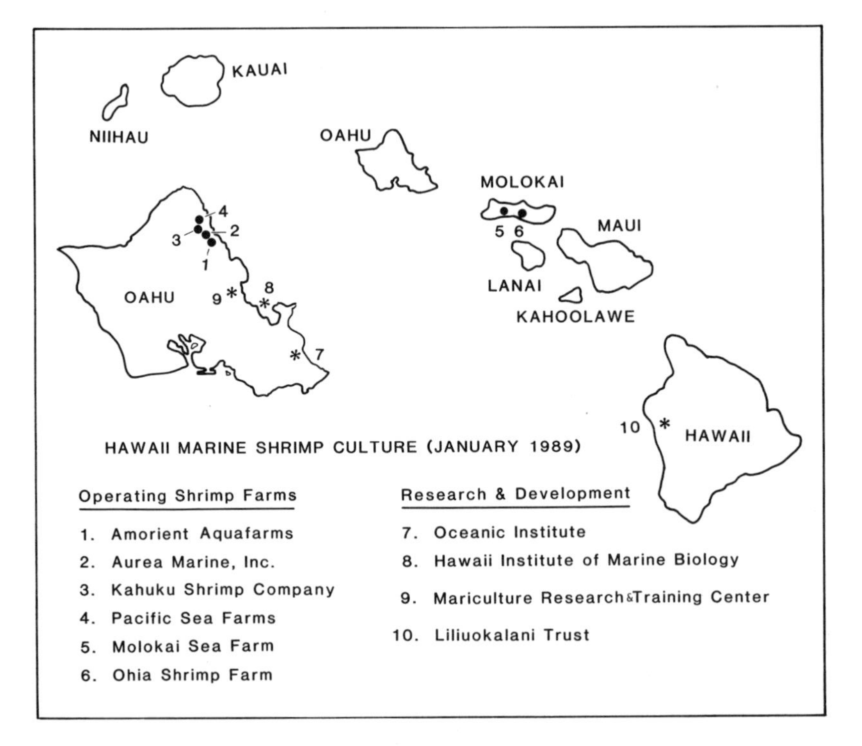

The locations of operating shrimp farms and research and development facilities in Hawaii are given in Figure 1.

RESEARCH ACTIVITIES AND SUPPORT

In 1988, research activities and support for marine shrimp aquaculture in Hawaii were as follows:

1) Center for Tropical and Subtropical Aquaculture - CSRS/ USDA Administered by the University of Hawaii (UH) and The Oceanic Institute (OI), Dr. Kevan Main, Executive Director.

Hawaii and Guam private sector representatives identified marine shrimp culture as a major area of interest. Priority funded projects were: wastewater effluent discharge (OI and UH), chemotherapeutics (UH and University of Arizona), and seed production (OI and Guam). These projects were funded at an estimated $500,000, divided about equally between Hawaii (UH and OI), the University of Arizona, and Guam.

2) U.S. Marine Shrimp Farming Program - CSRS/USDA. Administered by the Gulf Coast Research Laboratory and OI through the GCRL Consortium, Dr. Gary Pruder, Coordinator.

Consortium members are: the Gulf Coast Research Laboratory (GCRL), The Oceanic Institute (OI), Tufts University (TU), Texas Agricultural Experiment Station (TAES), Waddell Mariculture Center (WMC), and the University of Arizona (UA). The Consortium is dedicated to the high quality research, technology development and transfer necessary to expand the domestic marine shrimp farming industry.

The Hawaii portion of the shrimp farming program is executed by OI, Dr. James Wyban, Principal Investigator. The priority projects include development of intensive round pond growout, maturation, reproduction and hatchery technologies and economics. OI works in cooperation with all six commercial shrimp farms in Hawaii (Wyban et al. 1989, Deupree 1989, Moss 1989). Under the same program, OI is working with the State of Hawaii to establish a reliable source of disease-free seed and is developing a cooperative agreement with UH and Sea Grant to expand technology transfer activity.

OI marine shrimp farming program research activities were funded at $662,000 out of the CSRS/USDA total program award of $2,141,000 for FY88.

3) Aquaculture Development Program (ADP), Department of Land and Natural Resources, State of Hawaii, Mr. John Corbin, Director.

ADP functions in three areas: (1) statewide planning, coordination, and communication; (2) support services in permit acquisition, site selection, marketing and economics, and disease diagnosis and prevention; and (3) funding and co-funding research, development, and demonstration. It strives to create a pro-aquabusiness investment climate.

In 1988, ADP committed 42% of its $560,000 research, development, and demonstration project budget to marine shrimp. They are currently funding work on pond dynamics, shrimp growth response, reproductive endocrinology, shrimp pond nutrition, and virus detection. Projects are funded at the Mariculture Research and Training Center, the Hawaiian Institute of Marine Biology, the University of Hawaii (Manoa) and the University of Arizona.

4) Hawaii Office of Sea Grant - U.S. Department of Commerce, The Hawaii Sea Grant Program (UHSG), University of Hawaii, Dr. Jack Davidson, Director.

Only a small portion of the Sea Grant Budget is committed to aquaculture. In 1988, $150,000 to $200,000 was committed to marine shrimp research at the University of Hawaii. UHSG is concentrating its funding in areas amenable to controlled high quality research as compared to much less rigorous field research. Sea Grant (SG) and ADP develop a common research plan. SG generally commits its funding to the same areas identified under ADP above.

5) The Oceanic Institute Aquaculture Research Program - ARS/USDA, Dr. Eirik Duerr, Principal Investigator.

The Oceanic Institute, in cooperation with the Agricultural Research Service (ARS), carries out research projects dedicated to the incorporation of U.S. grains and oil seeds in aquaculture feeds (Duerr 1989). ARS investigators also carry out research projects at The Oceanic Institute. About 80% of the efforts expended by these two grants is directly applicable to marine shrimp hatchery and intensive growout feeds, or about $550,000 per year.

PRODUCTION ACTIVITIES AND INTERACTIONS

1) Amorient Aquafarms, Inc.
P.O. Box 131
Kahuku, Hawaii 96731 USA
(808) 293-8531
Dr. Linden Burzell, General Manager

At this time, Hawaii has only this one commercial company farming and selling marine shrimp in Hawaii. The farm consists of 143 one-acre ponds to produce in excess of 500,000 pounds of *Penaeus vannamei* and 50,000 pounds of *Macrobracium rosenbergii*. Eleven- to fifteen-gram shrimp are delivered to market five days a week. Their freshwater prawn culture utilizes seed produced and sold by the State of Hawaii. Amorient recently acquired an additional 34 1/4-acre ponds and one 1/2-acre pond which were originally part of Taylor Pryor's seafood plantation farm in Kahuku. These ponds are being developed as intensive ponds for both *P. vannamei* and *P. monodon*.

Amorient is a fully integrated shrimp farm with broodstock, maturation, reproduction, hatchery, and growout capabilities. Amorient sells surplus specific-pathogen-free broodstock to all markets and postlarvae to U.S. mainland and foreign marine shrimp farms.

Amorient currently receives premium prices for the shrimp produced. They are concerned about expanding marine shrimp production in Hawaii, as they believe the premium market for marine shrimp in Hawaii is very thin and added production could result in substantial reduction in selling price. The cost of producing shrimp in Hawaii is high. Amorient must have a premium market or the company will not survive.

Amorient is cooperating with The Oceanic Institute, Frito-Lay of Hawaii, and the East-West Research Institute to better define the price and market size for fresh large shrimp in Hawaii. Large shrimp, 26 count, have already been produced in a 15-week period in an OI round pond system. The market study will be completed soon.

In cooperation with The Oceanic Institute, Amorient is evaluating a 2,000 m^2 round pond shrimp growout system. The results of the evaluation to date were presented by Dr. Wyban (Wyban 1989). The pond yielded 21,000 pounds of shrimp in 46 weeks. Amorient is also cooperating with UH in an evaluation of oyster/shrimp co-culture under a project funded by the USDA, Dr. Jaw Kai Wang, Principal Investigator. The oyster growth results are reported by Lam et al. (1989) and Jakob et al. (1989).

2) Aurea Marine, Inc.
P.O. Box 38
Kahuku, Hawaii 96731 USA
(808) 293-1802
Mr. Terry Astro, President

Aurea is located on the site and uses some of the residual facilities of Tap Pryor's seafood plantation. Originally an oyster farm, Aurea first converted it to a tilapia farm in 1986. In the last few months, Mr. Astro has begun the cultivation of marine shrimp in one 1-acre concrete pond. The pond is currently stocked with *P. monodon*. Eight 1/4-acre concrete ponds are available for expansion. Aurea is completely dependent upon outside sources for postlarval seed, which is in very short supply. No market-sized animals have been produced. The lack of seed is the number one threat to Aurea's survival.

3) Kahuku Shrimp Company
P.O. Box 848
Kahuku, Hawaii 96731 USA
Mr. C. Bruce Smith, President

The Kahuku Shrimp Company (KSC) was started in the fall of 1988 by Mr. Smith, President and operator of the Kahuku Prawn Company. The prawn farm remains in operation, having produced between 30,000 to 40,000 pounds of *Macrobracium rosenbergii* in 1988. No marine shrimp have been produced to date.

KSC is envisioned as a fully integrated, 10-acre operation including maturation, hatchery, and intensive growout ponds, to produce large marine shrimp. The site was previously used by Marine Culture Enterprises for research, prior to their demise in 1988. Initial efforts are being expended to bring a hatchery on line and initiate growout.

KSC is cooperating with OI, which will provide initial supplies of specific-pathogen-free (SPF) nauplii. They consider the availability of SPF broodstock and the production of SPF seed critical to their success. KSC intends to make surplus supplies of postlarval shrimp available for sale to other Hawaii marine shrimp farms.

4) Molokai Seafarms, Inc.
P.O. Box 560
Kaunakakai, Hawaii 96748 USA
(808) 553-3547
Mr. Craig Blomquist, Principal

This marine shrimp farm was started in mid-1988 and has not yet produced shrimp for market. The site was formerly part of Orca Sea Farms. The facility consists of a maturation/hatchery facility and 18 acres of ponds. Molokai will utilize the semi-intensive culture system approach. Two ponds are currently stocked, one with *P. vannamei* and one with *P. monodon*. Further stocking is not possible because no postlarvae are available.

Molokai is considering a start-up of their hatchery using SPF nauplii to be supplied initially by OI. Adequate and reliable supplies of SPF postlarvae are their most urgent need.

5) Ohia Shrimp Farm Corp.
Star Route 61
P.O. Box 281
Kaunakakai, Hawaii 96748 USA
Greg Brown and Desmond Manaba, Partners

Ohia is a new facility with construction beginning in 1988. No shrimp have been marketed to date. Six 1/2-acre, 153'-diameter round ponds have been constructed. The ponds are 4.5' deep and have 10"-diameter center drains. Each two ponds drain into a central harvest box. Two additional ponds are provided for co-culture production using shrimp pond effluent.

Two of the large ponds have been stocked with *P. vannamei* from OI. No further marine shrimp stocking is possible until a source of postlarvae is developed. Ohia considers an adequate supply of postlarvae seed to be a number one priority.

6) Pacific Sea Farms
P.O. Box R
Laie, Hawaii 96762 USA
(808) 293-2466
Mr. James Ure, Farm Manager

Pacific Sea Farms took over the Marine Culture Enterprises production site. They have not yet produced shrimp for market. They expect to activate the MCE raceways and to expand into advanced shrimp culture systems they developed in the Cayman Islands. Pacific Sea Farms is modifying existing maturation and hatchery facilities to produce satisfactory yields of *P. vannamei*.

Pacific Sea Farms has also identified reliable and adequate supplies of specific-pathogen-free broodstock and postlarvae as a number one problem to be resolved.

INTERNATIONAL CONSULTING AND JOINT ACTIVITIES

Hawaii-based aquaculture consultants continue to work around the world, with special emphasis on marine shrimp, principally *P. monodon*.

1) Aquatic Farms, Ltd.
1164 Bishop Street, Suite 1608
Honolulu, Hawaii 96813 USA
(808) 531-8061
Dr. Edward Scura, President

Aquatic Farms employs a staff of 20 consultants and an administrative staff of five. The 1988 revenues were reported as approximately $2,200,000. They consulted in Indonesia, Burma, Pakistan, the Philippines, Thailand,

Honduras, Guatemala, Costa Rica, India, and China. Projects included site evaluation, production, marketing, and economic analysis. Aquatic Farms is increasing its participation in joint ventures in production, marketing, raw materials, and products.

2) Aquacultural Concepts
P.O. Box 560
Waimanalo, Hawaii 96795 USA
(808) 724-7975
Dr. Robert Shleser, President

3) Hawaii Aquaculture Company, Inc.
Kaimuki Technology Enterprise Center
1103 9th Avenue, Suite 206
Honolulu, Hawaii 96816 USA
(808) 733-2006
Dr. Spencer Malecha, President

These latter two companies are involved in aquaculture projects in South America, Central America, and/or the Caribbean. The percentage of their efforts in marine shrimp or the magnitude of revenues in Hawaii were not available.

A number of individual consultants from Hawaii are involved in international aquaculture projects.

SUMMARY

1) Approximately 500,000 pounds of marine shrimp and 100,000 pounds of freshwater prawns were cultured in Hawaii in 1988. The resulting revenues ranged from $2.5 to $3.5 million.

2) Marine shrimp rearing was undertaken by five new small farms in 1988. These farm operations were curtailed because of the lack of postlarvae. This problem is clearly the number one constraint to industry expansion at this time.

3) Federal support for Hawaiian marine shrimp research, exclusive of capital grants, was approximately $1.7 million in 1988.

4) State support for marine shrimp research to the University of Hawaii, the Mariculture Research and Training Center, and the Hawaii Institute of Marine Biology was approximately $250,000 in 1988.

5) International consulting generated approximately $2.5 million in revenues in 1988.

6) Total revenues from production, research, and consulting associated with marine shrimp was about $7.5 million in 1988 in Hawaii.

LITERATURE CITED

Corbin, J. and J. Sexton. 1984. Report of the Governor's aquaculture industry development committee. State of Hawaii, Aquaculture Development Program, Department of Land and Natural Resources, Honolulu, Hawaii, USA.

Deupree, R. 1989. Individual female *Penaeus vannamei* performance under long term maturation system conditions. Presented at Aquaculture '89, World Aquaculture Society, Los Angeles, California, USA, Feb. 1989.

Duerr, E. 1989. Outdoor evaluation of U.S. shrimp feeds for the intensive culture of *Penaeus vannamei*. Presented at Aquaculture '89, World Aquaculture Society, Los Angeles, California, USA, Feb. 1989.

Jakob, G. S., G. Pruder and J. K. Wang. 1989. Design and evaluation of an oyster growout tank utilizing shrimp pond effluent as feed. Presented at Aquaculture '89, World Aquaculture Society, Los Angeles, California, USA, Feb. 1989.

Lam, C. Y. and J. K. Wang. 1989. Growout trials of *Crassostrea virginica* in tanks fed shrimp pond water in Hawaii. Presented at Aquaculture '89, World Aquaculture Society, Los Angeles, California, USA, Feb. 1989.

Moss, S. 1989. Relative enhancement of *Penaeus vannamei* growth by selected fractions of intensive round shrimp pond effluent. Presented at Aquaculture '89, World Aquaculture Society, Los Angeles, California, USA, Feb. 1989.

Pruder, G., J. Wyban and J. Ogle. 1985. U.S. marine shrimp farming consortium: current status of domestic producers. The Oceanic Institute, Honolulu, Hawaii, USA.

Wyban, J., J. N. Sweeney and R. A. Kanna. 1989. Large shrimp production and carrying capacity in round ponds. Presented at Aquaculture '89, World Aquaculture Society, Los Angeles, California, USA, Feb. 1989.

MARINE SHRIMP AQUACULTURE IN MEXICO: CURRENT STATUS

Alejandro Flores Tom and Ernesto A. Garmendia Nunez

ABSTRACT

This paper presents a brief history of shrimp farming in Mexico, followed by a discussion of the main species cultured, the size and production capacities of the farms, the number of crops/yr, the magnitude of the harvests, the value of the crops, and major problems and perspectives related to the industry. The paper emphasizes commercial operations, and any omissions should be attributed to the authors. By December 1988, there were 100 registered shrimp farms in production in Mexico, totalling 7,288 ha of finished ponds, with an additional 2,215 ha under construction. Thus, a total of 9,503 ha may be available for operation in 1989. Seven hatcheries were producing in 1988, and seven more were under construction. The production capacity of the existing hatcheries is 149 million postlarvae/yr, and that of the new hatcheries will be 389 million postlarvae/yr.

BRIEF HISTORY OF SHRIMP AQUACULTURE IN MEXICO

The first shrimp farming trials in Mexico were conducted at the experimental facility of the Instituto Tecnologico y de Estudios Superiores de Monterrey in Guayamas, Sonora, in the early 1970's. Shortly thereafter, studies focused mainly on the intensive culture of the blue shrimp, *Penaeus stylirostris*, were begun in the state of Sonora in 1973 at the Puerto Penasco unit of the Centro de Investigacion Cientifica y Tecnologica de la Universidad de Sonora (CICTUS). Then, in 1977 the first shrimp farm was built in Sinaloa. It is located in the Los Carros inlet in front of the El Huizache marsh in the municipality of Mazatlan. The farm was designed to have 7.5 ha of semi-intensively operated ponds, and this is its current operating area.

In the early 1980's, four commercial shrimp farms were built in the state of Nayarit for the Sociedad Cooperative de Produccion Pesquera "Unica de pescadores Adolfo Lopez Mateos," S.C.L. At the beginning of 1985, CET del Mar La Paz in Baja California Sur started operating a hatchery and a small-scale (0.1 ha) experimental intensive round pond (CET del Mar 1985, 1986). Also, two more shrimp farms, with a total pond surface area of 328 ha, were built in 1985 in the state of Sinaloa. These were Las Grullas and Viveros de Camaron de Aqua Dulce, and they gave a strong impetus for the development of shrimp aquaculture in that state (Garmendia 1988). Finally, because of the growth of shrimp farming in Mexico, in October 1987 the Fisheries Ministry issued the national program for shrimp aquaculture through its general direction of aquaculture. This becomes the orienting and enhancing instrument for the emerging shrimp farming industry in Mexico. The growth of the industry in terms of pond area is summarized in Table 1, which shows a phenomenal 5.6-fold increase from 1985 through 1988.

Table 1. Area of operational shrimp grow-out ponds in Mexico by year (1985 through 1988) and the annual rate of increase.

YEAR	AREA (ha)	RATE OF INCREASE (%)
1985	1,249	—
1986	2,260	81
1987	5,338	136
1988	6,960	30

Source: Direccion General de Acuacultura, SEPESCA

COMMERCIAL SHRIMP AQUACULTURE IN MEXICO

Main Species

The three main species reared commercially in Mexico are *Penaeus californiensis*, *P. stylirostris*, and *P. vannamei*, with the white shrimp (*P. vannamei*) the most widely reared (Table 2). Three other species, all from

Table 2. Major species of marine shrimp cultured in Mexico, by coast and state.

STATE	SPECIES CULTURED	
	SCIENTIFIC NAME	COMMON NAME
PACIFIC COAST		
Baja California	*P. stylirostris*	Blue shrimp
Baja California Sur	*P. stylirostris*	Blue shrimp
	P. californiensis	Brown shrimp
Chiapas	*P. vannamei*	White shrimp
Jalisco	*P. vannamei*	White shrimp
Nayarit	*P. vannamei*	White shrimp
Sinaloa	*P. vannamei*	White shrimp
	P. stylirostris	Blue shrimp
Sonora	*P. stylirostris*	Blue shrimp
	P. vannamei	White shrimp
GULF OF MEXICO COAST		
Tamaulipas	*P. vannamei*	White shrimp
	P. aztecus	Brown shrimp
Campeche	*P. vannamei*	White shrimp

Source: Direccion General de Acuacultura, SEPESCA

the Gulf of Mexcio, are being studied at the research scale. These are *P. aztecus*, *P. setiferus*, and *P. duorarum*. The main objective of these studies is to develop culture technology for local species, thus avoiding the importation of non-native organisms and the potential risk of disease importation and spread. The concern over diseases may become an important problem to be faced by Mexico's developing shrimp aquaculture industry.

Size of the Industry

At the end of 1988, Mexico had 100 registered shrimp farms, with a total grow-out surface area of 7,288 ha (Table 3). Further, there were 49 farms with 2,215 ha in new ponds under construction. Of this amount, 643 ha was in expansion of 16 existing farms, while 1,572 ha were being built

Table 3. General technical information on operating shrimp farms in Mexico.

STATE	NUMBER OF FARMS	AREA (ha)	PRODUCTION CAPACITY (MT/yr)	NUMBER OF MEMBERS
Baja California	1	50	90	60
Baja Calif. Sur	3	21	61	90
Campeche	1	20	36	34
Chiapas	1	30	54	30
Jalisco	1	6	11	34
Nayarit	6	697	349	652
Sinaloa	75	6,075	5,493	4,328
Sonora	8	238	436	498
Tamaulipas	4	151	280	158
TOTAL	100	7,288	6,810	5,884

Source: Direccion General de Acuacultura, SEPESCA

on 33 new farms (Table 4). Thus, in 1989 Mexico was expected to have 9,503 ha of ponds for production of shrimp.

Farms in the Planning Stage

There is a close relationship between the number of projects submitted to the financial institutions for evaluation and the current development of commercial shrimp aquaculture. This statement is based upon the fact that the state of Sinaloa has 83% of the operating shrimp farms, 67% of the farms under construction, and 80% of the farms under evaluation.

The greatest development of shrimp aquaculture in Mexico is anticipated in the states of Sinaloa and Sonora. These two states are expected to eventually have 97,826 ha (92%) of the total 106,046 ha of ponds projected for the country (Table 5). The status of the planned projects varies from integrated technical, economic and financial viability to pre-feasibility investment studies.

Table 4. General information on shrimp farms under construction in Mexico.

STATE	EXPANSION OF EXISTING FARMS		NEW FARMS	
	No.	Ha	No.	Ha
Baja California Sur	-	-	1	15
Campeche	-	-	1	10
Chiapas	1	40	2	100
Nayarit	-	-	4	240
Sinaloa	12	453	19	1,037
Sonora	3	150	1	5
Tamaulipas	-	-	3	115
Veracruz	-	-	1	30
Yucatan	-	-	1	20
TOTAL	16	643	33	1,572

Source: Direccion General de Acuacultura, SEPESCA

Types of Shrimp Farms

The three major types of farms - extensive, semi-intensive, and intensive - are all found in Mexico. Semi-intensive operations dominate the Mexican industry in terms of number of farms and total production, while the greatest amount of pond area is in extensive culture and the least in intensive (Table 6). The average sizes of farms by production type are as follows: extensive, 95.85 ha; semi-intensive, 57.18 ha; and intensive, 5.0 ha.

Total Production and Number of Crops/Year

Due to construction and operational problems, 6,960 (95.5%) of the 7,288 ha registered in 1988 were actually in production that year, yielding a total of 3,200 metric tons. Approximately 15% of the farms produced two crops/yr. The other 85%, which were limited to one crop, were either (1) extensive farms dependent upon wide tidal fluctuations which makes them able to operate only 6-8 months of the year and/or (2) relying on the avail-

Table 5. Planned development of shrimp farms in Mexico, by state.

STATE	NUMBER OF FARMS	AREA (ha)	PERCENT OF POTENTIAL
Baja California	2	800	0.75
Baja California Sur	2	250	0.24
Campeche	2	340	0.32
Chiapas	10	1,910	1.80
Nayarit	3	2,200	2.07
Oaxaca	9	1,890	1.78
Sinaloa	306	85,024	80.18
Sonora	46	12,802	12.07
Tamaulipas	3	330	0.31
Veracruz	2	500	0.47
TOTAL	385	106,046	100.00

Source: Direccion General de Acuacultura, SEPESCA

ability of wild postlarvae to stock the ponds. The wild seed is most abundant from June to October (5 months) and in the northwest, which is where >90% of the infrastructure for the shrimp farming industry is found (Juarez and Garmendia 1988). It is also important to note that in 1988 only seven farms used hatchery-reared postlarvae to partially or completely stock their grow-out facilities. In two of these cases, the postlarvae were shipped in from outside Mexico.

Average Production (kg Whole Shrimp/ha/Crop)

Average yields achieved by the various shrimp farms, depending on the culture approach followed, were as follows:

Extensive culture: one grow-out cycle, with an average yield of 335 kg heads-on shrimp or 214 kg tails/ha/yr.

Semi-intensive culture: one crop; average yield of 650 kg heads-on shrimp or 416 kg tails/ha/yr. Some farms reported two crops, with >1,500 kg heads-on shrimp/ha/yr.

Table 6. Existing shrimp farms in Mexico, summarized by state and culture system utilized.

STATE	TOTAL FARMS		NO. AND AREA BY CULTURE SYSTEM EXTENSIVE		SEMI-INTEN.		INTENSIVE	
	No.	Ha	No.	Ha	No.	Ha	No.	Ha
Baja California	1	50	-	-	1	50	-	-
Baja Calif. Sur	3	21	-	-	3	21	1	15
Campeche	1	20	-	-	1	20	-	-
Chiapas	1	30	-	-	1	30	-	-
Jalisco	1	6	-	-	1	6	-	-
Nayarit	6	697	2	532	4	165	-	-
Sinaloa	75	6,075	44	3,877	30	2,195	1	3
Sonora	8	238	-	-	7	237	1	1
Tamaulipas	4	151	-	-	3	150	1	1
TOTAL	100	7,288	46	4,409	50	2,859	4	20

Source: Direccion General de Acuacultura, SEPESCA

Intensive culture: 1988 production of 100 MT of heads-on shrimp; average yield of 5,000 kg heads-on shrimp or 3,200 kg tails/ha/yr.

Shrimp Size at Harvest

The size of the product at harvest varied depending on the duration of the grow-out cycle and the type of culture approach.

The average size of the shrimp harvested from the intensive farms was rather small (61-70 count or smaller), due to the high rearing density and to operational problems near the end of the grow-out cycles. The operational problems were mainly caused by a breakdown of the aeration system and inadequacy of financial resources during the project's experimental stage (Juarez and Garmendia 1988).

The semi-intensive farms produced a wide variety of sizes, due to different stocking densities used. Harvest sizes ranged from 21-25 count to 61-

70 counts, with 36-40 and 41-50 count sizes dominant (Juarez and Garmendia 1988).

The largest shrimp were harvested from the extensive farms. These operations had grow-out cycles of approximately eight months, low stocking densities (about 3 shrimp/m^2), and a mean survival of >70%. The product was mainly 31-35 and 26-30 count. Some extensive farms produced smaller shrimp and lower yields as a result of higher stocking density and high mortalities during the last months of the grow-out cycle (Juarez and Garmendia 1988).

Crop Value

Marine shrimp farms contributed 3,200 MT of heads-on shrimp in 1988, amounting to 4.1% of Mexico's total shrimp production. About 80% of the 1988 aquaculture production was exported to the US market, with an estimated average price of US$4.50/lb of tails. The export volume was estimated to be 3,604,480 lbs with a total value of US$16,220,000.

Hatcheries

There are seven penaeid shrimp hatcheries in operation in Mexico; six are located on the Pacific coast and one on the Gulf of Mexico. Together, they have a production capacity of 149 million postlarvae/yr (Table 7). The primary species produced in these hatcheries is *P. sytlirostris*, due to the influence of the technology derived from the work done at CICTUS of the University of Sonora at Puerto Penasco.

In addition, seven new hatcheries are under construction, with six again on the Pacific coast and one on the Gulf of Mexico. These hatcheries are expected to have a production capacity of 389 million postlarvae/yr (Table 8). Most of these hatcheries will produce *P. vannamei*, based on experience gained by Ecuadorian and Panamanian hatcheries. The new hatcheries will equal the existing ones in number, but will more than double the current production capacity.

Table 7. Operating penaeid shrimp hatcheries in Mexico.

STATE	HATCHERY NAME	PRODUCTION CAPACITY (10^6 PLs/yr)	SPECIES
Baja Calif. Sur	CET DEL MAR LA PAZ	24	*P. califoriensis*
Jalisco	BARRA NAVIDAD UAG	2	*P. vannamei*
Sinaloa	UNIVERSIDAD AUTONOMA DE SINALOA	18	*P. vannamei* *P. stylirostris*
Sonora	BIOTECMAR	30	*P. stylirostris*
	CICTUS	30	*P. stylirostris*
	AQUACULTIVOS DEL PACIFICO	30	*P. stylirostris*
Tamaulipas	UNIDAD MARINA UAT	15	*P. vannamei*
TOTAL	7	149	

Table 8. Marine shrimp hatcheries under construction in Mexico.

STATE	HATCHERY NAME	PRODUCTION CAPACITY (10^6 PLs/yr)	SPECIES
Baja Calif. Sur	AQUACULTORES DE LA PENINSULA	50	*P. stylirostris* *P. californiensis*
Campeche	ACUACULTORA CAMPECHANA	36	*P. vannamei*
Colima	TECUANILLO	36	*P. vannamei*
Nayarit	SAN BLAS	15	*P. vannamei*
Sinaloa	FRSCIP CENTRO SINALOA	180	*P. vannamei*
	CAMARICULTORES DE SINALOA	36	*P. vannamei* *P. stylirostris*
Sonora	CULTIVOS DEL MAR DE CORTEZ	36	*P. vannamei*
TOTAL	7	389	

Source(Tables 7 & 8): Direccion General de Acuacultura, SEPESCA

RESEARCH AND EXTENSION FOR SHRIMP AQUACULTURE DEVELOPMENT IN MEXICO

It is necessary to link the research area with production if Mexico is to rapidly take advantage of new advances in technology. For 1988, 66 research projects were registered, of which 44 have been concluded and 22 are under development. The programs involved 59 investigators from 25 institutions, and focused mainly on the need for technical and scientific information for the rational use of resources in shrimp aquaculture and for the improvement of biotechnologies leading to increases in productivity/ha. Special emphasis has been placed on the evaluation of occurrence and availability of wild larvae, postlarvae and juveniles, on studies to integrate basic ecosystem information, and on bioassays which adapt and improve the methodologies for shrimp larval rearing in hatcheries and grow-out in ponds (Anonymous 1987).

Extension is primarily the responsibility of public institutions. Through 1988, there have been nine training courses in Mexico which trained 326 technicians. Seven of the courses were oriented toward grow-out operations and were attended by 296 trainees. The other two courses, which focused on hatchery methods, were attended by 30 technicians. CET del Mar has contributed to the extension activity, developing five marine shrimp courses for 74 aquaculture technicians from different institutions and producing 71 marine aquaculture specialists in five classes through its aquaculture education/training program begun in 1982.

The level of the courses presented in Mexico is constantly improving. The 1988 events included foreign speakers, mainly from the Republic of Panama (Anonymous 1987, 1988), and for 1989 international shrimp culture training courses are planned.

MAIN PROBLEMS

Mexico's shrimp farms are consolidating technologically. Individually they are achieving increasing yields, and the nation's yield average for each of the culture approaches is improving. However, to a great extent, improvement of the operating units depends on the supply of seed, either from the wild or from hatcheries. It is perhaps paradoxical that in the state of Sinaloa, where 80% of the grow-out pond area is located, there are no operating hatcheries. Further, postlarvae are available in the wild only for a short period relative to the industry needs. As a result, many farms can accomplish only one grow-out cycle/yr (Juarez and Garmendia 1988).

Some of the production units exhibit construction deficiencies because of lack of experience in the field. This situation applies primarily to the projects that were developed first. However, subsequent projects have profited from lessons learned in the earlier ones. Also, the technological approaches being applied in Mexico have been taken from the experiences of countries like Ecuador and Panama and adapted to the local situations (Garmendia et al. 1987).

Marine shrimp hatcheries have not yet consolidated their commercial activities. This affects the availability of postlarvae for the farms. The quality of feed and other materials for grow-out appear to be acceptable, but it is believed that better quality feeds may improve yields.

Undoubtedly, one of the significant limiting factors for the development of shrimp aquaculture is financing. The problem is twofold; the technical and financial proposals presented to the banks lack necessary information while the investment requested from several potential producers exceeds the financial capabilities of the banks.

The subject of disease must also be seriously considered, even if it has not yet had a serious detrimental effect on the development of the industry. Of special concern must be the transportation and introduction of non-native species to new environments, and the potential ecological risks of such introductions. At present, the Fisheries Ministry is conducting studies in order to solve the existing problems, and is involving the education/research and financial institutions as well as the producers.

FUTURE PERSPECTIVES

The future for marine shrimp culture in Mexico appears encouraging due to the rapid expansion of shrimp farms and hatcheries on both coasts. For the period 1985-1988, the annual average growth rate in grow-out pond area for shrimp was 69%. This is a superior growth rate when compared to other aspects of the nation's economy, and similar to growth rates observed in countries that are currently the leaders of the shrimp aquaculture industry. The current situation of the marine shrimp aquaculture activities in Mexico reflects a continuously growing industry which, even if it has not yet reached the development level stated in the national program for shrimp aquaculture, has achieved important advances and applied many world-wide experiences which will accelerate and consolidate its development. It is interesting to note that, from an economic point of view, a high percentage of the production units are not the only sources of income for the producers but are supplementary to their incomes.

The first hatcheries designed and built in Mexico were projected for small and medium scale production. They basically produce for their own grow-out units and have limited capacity as suppliers for other farms. However, the hatcheries which are under construction have a higher production capacity, which should improve the availability of postlarvae for the coming grow-out cycles.

It is anticipated that, in the near future, there will be increased production from shrimp aquaculture in Mexico, both from the addition of more ha of grow-out ponds as well as from increases in productivity of existing ponds. This latter will result from a larger supply of postlarvae for a second stocking of ponds each year, and from better management and feeding plus improvements to the production units themselves. As a result, many of the extensive type operations built in 1986 are tending to become semi-intensive farms.

LITERATURE CITED

Anonymous. 1987. Programa Nacional de Cultivo de Camaron. Direccion General de Acuacultura, Secreteria de Pesca, Mexico.

Anonymous. 1988. Informe de avance al 31 de Julio de 1988 del Programa Nacional de Cultivo de Camaron. Direccion General de Acuacultura, Secretaria de Pesca, Mexico.

CET del Mar. 1985. Primera reunion de resultados del proyecto desarrollo experimental de cultivo de camaron. Ed. CET del Mar La Paz. B. C. S. 47 pp.

CET del Mar. 1986. Analysis on the first operation year of shrimp culture facilities in La Paz, Mexico. Presented at the GTL US/Mexican Marine Symposium, San Bernadino, CA. 30 pp.

Garmendia, N. E. A. 1988. Situacion actual y perspectivas del cultivo de camaron en Mexico. Secretaria de Pesca, Mexcio.

Garmendia, N. E. A., M. R. Lopez, A. G. Ochoa, S. A. Pares, and R. H. Ramirez. 1987. Informe de la comision realizada por tecnicos mexicanos a la Republica de Ecuador. Secretaria de Pesca, Mexico.

Juarez, P. J. R. and N. E. A. Garmendia. 1988. Evaluacion de la operacion de 10 granjas comaroneras en Mexico. Secretaria de Pesca, Mexico.

PART II:

TECHNOLOGICAL ADVANCES

FURTHER INTENSIFICATION OF POND SHRIMP CULTURE IN SOUTH CAROLINA

Paul A. Sandifer, Alvin D. Stokes and J. Stephen Hopkins

ABSTRACT

Penaeus vannamei were reared intensively in 0.25 ha ponds at the Waddell Mariculture Center. Duplicate ponds were stocked with 100 and 200 postlarvae/m^2 and reared for 182-188 days. Nursery-reared juveniles (1.9 g) were stocked at 40 shrimp/m^2 in a 0.1 ha pond. This pond yielded 9,007 kg/ha of 24.3 g shrimp at harvest after 172 days. Mean survival, production and weight at harvest for shrimp stocked as postlarvae at 100/m^2 were 49.2%, 11,065 kg/ha and 22.7 g, while at the 200/m^2 density these averages were 58.7%, 21,300 kg/ha and 18.8 g. One replicate at the 200 shrimp/m^2 density produced 25,919 kg/ha. The relatively low survival rates reflected dissolved oxygen problems experienced over the growing season.

INTRODUCTION

The growing season for shrimp in South Carolina and other parts of the southern United States is climatically limited to five-seven months. To be competitive with shrimp growers in more tropical parts of the world and to recoup capital costs, South Carolina farmers must maximize production per unit of farm areas during this restricted growing season. In addition, most sites suitable for shrimp farming in South Carolina are no larger than a hundred ha at best (most are considerably smaller). This situation precludes the building of very large, extensive-type shrimp farms. Instead, what is needed for South Carolina and other areas of the US is a highly efficient, intensive cultivation system that does not require a large land area to be commercially viable. Thus, intensification of pond production of shrimp has been a major focus of the Waddell Mariculture Center shrimp research program since 1985.

In previous studies (Sandifer et al. 1987, 1988 in press), we have demonstrated the following:

(1) Mean size at harvest was little affected by stocking density within the range 20-200 animals/m^2;

(2) Survival declined with increasing stocking density, reflecting the need for improved management at the higher densities;

(3) Standing crop biomass at harvest increased directly with stocking density, reaching a maximum of 12.7 MT/ha/crop in 1987.

The purpose of the present paper is report our most recent results from intensification experiments conducted during 1988 and place these into context with our previous data and results obtained elsewhere. The 1988 studies had two primary objectives: (a) to evaluate the effect of stocking nursery-reared juveniles and (b) to compare yields from very high stocking densities ($\geq$ 100 postlarvae/m^2).

MATERIALS AND METHODS

All work reported here was conducted at the James M. Waddell, Jr. Mariculture Research and Development Center located at Bluffton, SC. This Center is a field station of the Marine Resources Division of the South Carolina Wildlife and Marine Resources Department and is also affiliated with the South Carolina Agriculture Experiment Station system (Clemson University).

The studies involved four 0.25 and one 0.1 ha ponds. These are earthen ponds lined with 1 mm thick high density polyethylene to prevent seepage. The pond floor is covered with a 26-cm deep earth layer to provide natural bottom conditions.

Objectives of the 1988 studies were to:

(1) Compare performance of nursery-reared juveniles and direct-stocked postlarvae with respect to survival, growth, and crop yield; and

(2) evaluate very high stocking densities of 100 and 200 postlarvae/m^2 in replicate ponds.

With regard to the latter objective, we anticipated defining critical standing crop and/or carrying capacity (an upper limit of intensification) for this system.

Prior to stocking, all experimental ponds were allowed to dry. Soil pH levels were 6-6.5, so the ponds were limed at rates of approximately 1,000-1,500 kg/ha (Table 1). Fertilization to induce phytoplankton blooms was via a combination of 544 kg/ha of cottonseed meal, 544 kg/ha alfalfa, and 2.5-5 l/ha of 10-34-0 liquid fertilizer (Table 1).

Table 1. Soil treatment and pond fertilization for intensive pond culture of *Penaeus vannamei* in South Carolina, 1988

Pond No.	Soil pH	Liming Rate (kg/ha)	Fertilization Rate Alfalfa (kg/ha)	C.S.M.* (kg/ha)	10-34-0 Liquid (l/ha)
S1	—	1,587	—	—	2.5
M3	6.5	1,088	544	544	5.0
M6	6.3	1,088	544	544	5.0
M4·	6.4	1,088	544	544	5.0
M9	6.1	1,088	544	544	5.0

* cotton seed meal

All shrimp utilized were *Penaeus vannamei* obtained as postlarvae from Amorient Aquafarm (Hawaii). An initial shipment of 800,000 postlarvae was stocked in a 0.1 ha nursery pond on 3 March 1988. This pond was subsequently enclosed with a plastic cover inflated by a small blower. The plastic cover maintained water temperatures in the covered pond an average of 6.5 ± 1.4 C higher than those in an uncovered, control pond from late March through April 1988. Juvenile shrimp were harvested by draining on 29 April 1988 and were subsequently stocked into a 0.1 ha pond at 40 shrimp/m^2. Growth and survival of the juveniles were compared with ponds stocked with postlarvae for a diet evaluation (Hopkins et al. 1989).

Temperature and dissolved oxygen concentration were recorded daily at sunrise. Dissolved oxygen level was also generally checked at night as needed during the latter half of the growing season. Salinity and pH were measured once or twice per week. The shrimp were seine-sampled to determine growth rate and adjust feeding once/week.

All ponds were fed Zeigler 40% protein Shrimp Grower ration. The diet was stored in bulk on site and distributed to the ponds via a commercial feed blower. The shrimp were fed equal amounts three times/day, at 0800, 1700 and 2400 hrs, except one or more feedings were reduced or eliminated when sunrise dissolved oxygen levels were < 2.5 mg/l. The maximum feed rate attained in one of the 200 shrimp/m² ponds was 165 kg feed/ha/day, but most of the time feeding did not exceed 100 kg/ha/day.

One hp electric paddlewheel aerators were used in all ponds. Aeration rates were established prior to stocking, based on experience. The juvenile pond received 10 hp/ha of aeration, while the 0.25 ha ponds stocked with 100 and 200 postlarvae/m² received 12 and 24 hp/ha, respectively.

Routine water exchange appears to be an important requirement for successful intensive shrimp culture. Like aeration, exchange rates were based on experience, along with daily water quality observations. In this study, cumulative water exchange rates ranged from 16-23%/day, with a maximum daily rate of 50% for a short period (Table 2). Exchange rate was generally increased slightly with increasing stocking density.

Table 2. Water exchange rates for intensive culture of *Penaeus vannamei* in South Carolina, 1988, (PL=postlarvae; J=juvenile).

Stocking Density (No./m²)	Seed Size (PL or J)	Maximum Exchange Rate (%/day)	Overall Average Exchange Rate (%/day)
40	J	30	23.3
40	PL	50*	15.8
100	PL	40	17.2
200	PL	35	19.1

*1 pond had an exchange rate of 50% for three days due to a specific DO problem; otherwise, maximum exchange rate was 25%/day.

RESULTS

Water Quality Observations

Mean dissolved oxygen concentrations ranged from 4.2-4.8 mg/l (Table 3). Surprisingly, the poorest dissolved oxygen conditions (with minimums of 0.5 and 0.8 mg/l for the two replicates, respectively) occurred in the 100 postlarvae/m² ponds. Temperatures averaged 25.5-26 C, with minimum and maximums of 14.2 and 32.2 C, respectively. Minimum temperatures occurred at the end of the growing season, just before harvest. Salinity was similar in all ponds, ranging from 22-32 ppt with a mean (± standard deviation) of 29 ± 2 ppt.

TABLE 3. Summary of water quality observations during intensive pond culture of *Penaeus vannamei* at the Waddell Mariculture Center, South Carolina, 1988. Data are given as means ± standard deviation and range for daily observations over the entire growing season.

POND SIZE (ha)	STOCKING DENSITY (No/m²)	DISSOLVED OXYGEN (mg/l)		TEMPERATURE (C)	
		Mean ± S.D.	Range	Mean ± S.D.	Range
0.1	40*	4.3 ± 1.1	2.3 - 8.5	26.1 ± 2.7	17.0-31.2
0.25	100	4.6 ± 1.5	0.5 -11.5	25.5 ± 3.2	16.5-31.8
0.25	100	4.2 ± 1.6	0.8 -11.8	25.8 ± 3.2	17.0-32.2
0.25	200	4.7 ± 1.4	1.0 -10.2	25.3 ± 3.3	14.2-31.5
0.25	200	4.8 ± 1.1	1.1 - 9.7	25.6 ± 3.0	16.5-31.0

* Juvenile shrimp, rather than postlarvae; other densities were stocked with postlarvae.

Yield from Nursery-Reared Juveniles

After transfer to the growout pond, growth and survival of the nursery-reared juveniles were excellent (Table 4). Results for the nursery-reared animals were compared with those for direct-stocked postlarvae which were involved in a feeding trial (Hopkins et al. 1989). Note that logistics forced a four-week difference in rearing period for the two groups. Nevertheless,

Table 4. Effects of stocking nursery reared juveniles on production of *Penaeus vannamei* in South Carolina, 1988 (stocking density was $40/m^2$).

Pond Size (ha)	N	Duration (days)	Survival (%)	Initial Wt. (g)	Harvest Wt. (g)	Production (kg/ha)
0.1	1	172	92.9	1.9	24.3	9,007
0.25	4	144	89.0	0.1	15.2	5,511

even if the mean size and biomass at harvest were adjusted upward in the postlarvae ponds to account for this difference in growing period, yield from the juvenile treatment would still be appreciably greater. Here we harvested 9 MT/ha of 24 g shrimp. Our previous highest single pond yield at this density was 7.5 MT/ha of 18 g shrimp (Sandifer et al. 1987).

Very High Density Rearing Trials

Table 5 summarizes the production data for postlarvae stocked at 100 animals/m^2. Survival was relatively low, averaging 49% and reflecting the dissolved oxygen problems these ponds experienced. Growth rate averaged 1.1 ± 0.4 g/week for weeks 6 through 25 (that is, after the first 5 weeks of essentially nursery growth), with excellent mean size at harvest (22.7 g). Mean harvest yield of 11 MT/ha/crop was less than that (12.7 MT/ha) obtained in a 0.1 ha pond in 1987 (Sandifer et al. 1988). Feed conversion averaged 2.8 compared to 2.1 in 1987. However, mean size in 1988 was 9 g greater than that obtained in 1987. The primary reason for the differences in size, harvest biomass and feed conversion between the two years was the difference in survival. Survival in 1987 was 94%, while that in 1988 was only 49%.

Production data for the 200 postlarvae/m^2 ponds are summarized in Table 6. Survival was low (40%) in one replicate due to an acute dissolved oxygen problem, but survival in the other was excellent (77%). Mean shrimp size at harvest (18.8 g) was roughly 4 g less than that observed at the 100 shrimp/m^2 density. Mean growth rate for weeks 6 through 22 was 1.1 ± 0.5 g/wk, essentially the same as that observed at 100 shrimp/m^2. Harvest yield averaged 21.3 MT/ha/crop, with maximum yield of 25.9 MT/ha and a mean food conversion ratio of 2.5.

Table 5. Growth and production of *Penaeus vannamei* reared in duplicate 0.25 ha ponds stocked at 100 postlarvae/m² at the Waddell Mariculture Center, South Carolina, 1988.

	Rep. 1	Rep. 2	Mean
Initial Mean Weight (g)	0.001	0.001	0.001
Harvest Mean Weight (g)	24.0	21.3	22.7
Duration (days)	186	188	187
Survival (%)	44.1	54.2	49.2
Production (kg/ha)	10,581	11,549	11,065
Feed Conversion	2.89	2.74	2.82

Table 6. Growth and production of *Penaeus vannamei* reared in duplicate 0.25 ha ponds stocked at 200 postlarvae/m² at the Waddell Mariculture Center, South Carolina, 1988.

	Rep. 1	Rep. 2	Mean
Initial Mean Weight (g)	0.001	0.001	0.001
Harvest Mean Weight (g)	20.7	16.8	18.8
Duration (days)	187	182	184.5
Survival (%)	40.3	77.0	58.7
Production (kg/ha)	16,681	25,919	21,300
Feed Conversion	3.03	2.02	2.53

Comparison of mean growth curves for the 100 and 200 postlarvae/m² densities show little difference in slope (i.e., growth rate) (Fig. 1). It appears that the difference in mean sizes apparent at harvest was the result of a slowing of growth in the 200/m² ponds during the last few weeks of the rearing period when biomass was highest and pond temperatures lowest.

Similarly, the growth curve for the 40/m² juvenile pond exhibited essentially the same slope as that observed for the higher density ponds stocked with postlarvae. Weekly mean growth rates were similar and variable for both the 100 and 200/m² densities (Fig. 2). During some weeks at both densities observed growth rates exceeded 1.8 g/wk.

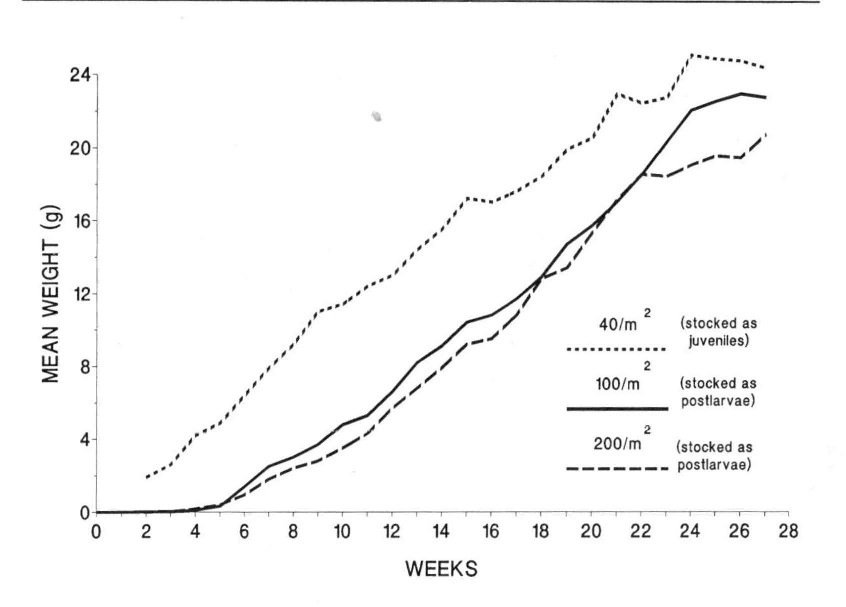

Figure 1. Comparison of average growth curves for *Penaeus vannamei* stocked at 40 juveniles/m² and 100 and 200 postlarvae/m² in South Carolina, 1988.

DISCUSSION

Neither critical standing crop nor carrying capacity for the Waddell Center pond system were determined. Even at a stocking density of 200 postlarvae/m² and harvest yield to 25.9 MT/ha, there was no indication that carrying capacity was being approached. It is obvious, though, that higher aeration rates are required to maintain suitable dissolved oxygen concentrations at these densities. Growth rate did slow at this density near the end of the growing period, probably as a result of the combination of high standing crop biomass and lowered water temperatures. Further work is needed to determine the upper production limit for intensive pond culture of shrimp. Data from the present study indicate that carrying capacity of the culture system described here exceeds 25 MT/ha. Further, the production level of 25.9 MT/ha recorded here is, to our knowledge, the highest production yet achieved anywhere in pond culture of shrimp of any species (see, for example, Chen et al. 1987, 1988; Chien et al. 1987, 1989; Wyban and Sweeney 1989; Wyban et al. 1988).

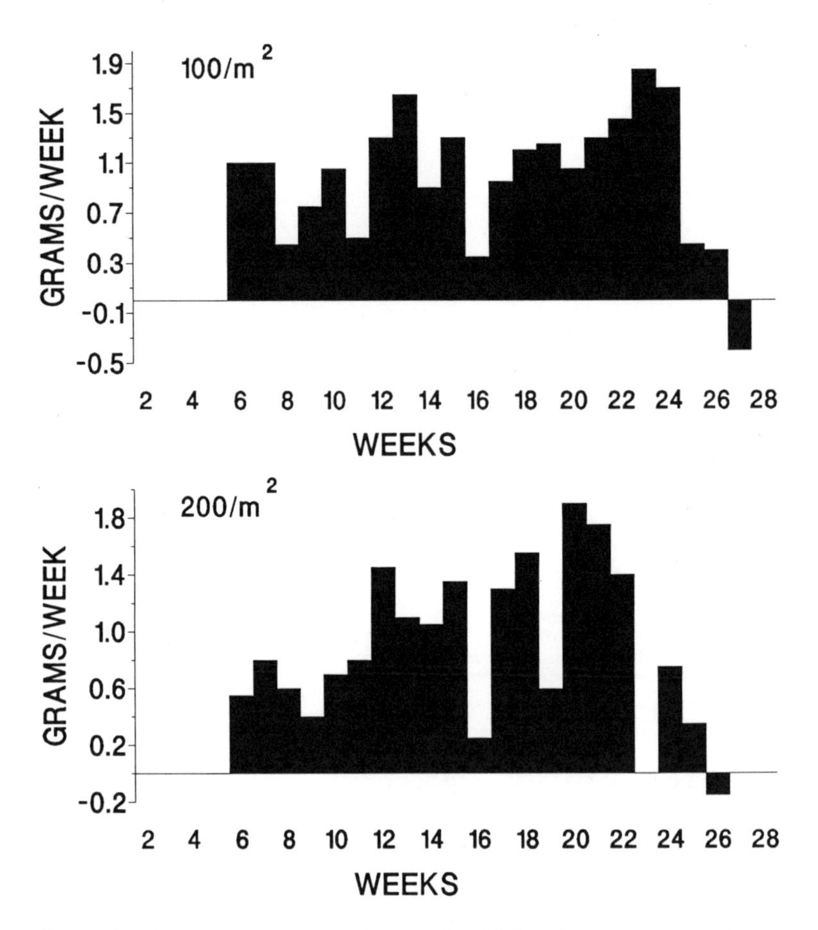

Figure 2. Average weekly growth rates (g/wk) for *Penaeus vannamei* reared at stocking densities of 100 and 200 postlarvae/m^2 in South Carolina, 1988.

Addition of these results to data from our previous studies (Table 7) reconfirmed our findings that standing crop biomass at harvest increases directly with stocking density (Sandifer et al. 1987, 1988, in press). Also, survival tended to decrease with increasing density, reflecting the increasing pond management needs. We believe that survival can be substantially improved at the higher densities as more experience is gained in working at

Table 7. Summary of production results (means ± standard deviation) for *Penaeus vannamei* stocked as postlarvae (0.001 - 0.1g) at different densities in earthen ponds at the Waddell Mariculture Center, 1985-1988. (N = number of ponds at that density, regardless of pond size).

Stocking Density (No/m²)	Number of Observations (N)	Mean Weight (g)	Survival (%)	Harvest Biomass (kg/ha)
12	3	19.7 + 0.9	100*	2,477 + 98
20	3	13.4 + 0.7	92.4 + 2.9	2,487 + 182
40	15	15.4 + 1.9	86.1 + 9.6	5,331 + 837
60	5	13.7 + 2.8	78.7 + 18.3	6,202 + 702
100	3	19.6 + 5.5	64.1 + 26.4	11,603 + 1,051
200	2	18.8	58.7	21,300

* Survival rates were 99-114%, indicating overstocking.

these production levels. No clear trend in mean weight versus stocking density is evident. Other factors, such as annual climatic variations, survival, and particularly seed stock source differences among years, appeared to affect size more than stocking density.

Data presented here and previously (Sandifer et al. 1987, 1988, in press) show that average growth rates of ± 1 g/wk can be expected routinely for *P. vannamei* reared intensively in ponds in South Carolina. We have observed much higher growth rates (> 1.8 g/wk) for short periods, but have not been able to maintain growth of > 1.5 g/wk over a growing cycle as did Wyban and Sweeney (1989). This deserves additional study. For example, if average growth rates could be increased to 1.5 g/wk, mean size at harvest should reach 30 to 37 g, depending on whether postlarvae or nursery-reared juveniles were stocked initially.

The Waddell Center experiments also indicate that if postlarvae are stocked directly into growout ponds (that is, the 4-6 week nursery period is completed in the growout pond), then about the largest mean shrimp size likely to be obtained at harvest is 19-20 g. Many harvests are likely to yield smaller shrimp. If instead the nursery period is completed before the beginning of the outdoor rearing period and larger juveniles, rather than postlarvae are stocked, then harvest size can be expected to average perhaps 24-25 g. Since the value of the harvest depends in large part on the size of the shrimp, this difference can be significant.

LITERATURE CITED

Chen, J. C., P. C. Lin, Y. T. Lin and C. K. Lee. 1987. Highly- intensive culture study of tiger prawn *Penaeus monodon* in Taiwan. *in* N. De Pauw, E. Jasper, H. Ackefors, and N. Wilkins, editors. Proceedings of Aquaculture Europe 1987. European Aquaculture Society, Bredene, Belgium.

Chen, J. C., P. C. Liu, Y. T. Lin, and C. K. Lee. 1988. Super intensive culture of red-tailed shrimp *Penaeus penicillatus*. Journal of the World Aquaculture Society 19(3): 127-131.

Chien, Y. H., I. C. Liao, and C. M. Yang. 1987. The evolution of prawn grow-out systems and their management in Taiwan. IChemE Symposium Series No. 111.

Chien, J. C., P. C. Liu and Y. T. Lin. 1989. Culture of *Penaeus monodon* in an intensified system in Taiwan. Aquaculture 77(4): 319-328.

Hopkins, J. S., A. D. Stokes and P. A. Sandifer. 1989. Effectof three feeding regimes on production of white shrimp, *Penaeus vannamei*, in intensive pond culture. Presented at Aquaculture '89, World Aquaculture Society, Los Angeles, CA, February 1989, unpublished.

Sandifer, P. A., J. S. Hopkins, and A. D. Stokes. 1987. Intensive culture potential of *Penaeus vannamei*. Journal of the World Aquaculture Society 18(2): 94-100.

Sandifer, P. A., J. S. Hopkins, and A. D. Stokes. 1988. Intensification of shrimp culture in earthen ponds in South Carolina: progress and prospects. Journal of the World Aquaculture Society 19(4): 218-226.

Sandifer, P. A., J. S. Hopkins, A. D. Stokes, and G. D. Pruder. In press. Technological advances in intensive pond culture of shrimp in the United States. Proceedings, Frontiers in Shrimp Research Symposium, Washington, DC, June 1988, Elsevier Press.

Wyban, J. A. and J. N. Sweeney. 1989. Intensive shrimp growout trials in a round pond. Aquaculture 76: 215-225.

REASSESSMENT OF THE PROSPECTS FOR FRESHWATER PRAWN CULTURE IN THE UNITED STATES: COMPLEMENTARY RESEARCH EFFORTS IN HAWAII AND MISSISSIPPI*

Louis R. D'Abramo, Spencer R. Malecha, Marty J. Fuller, William H. Daniels, and John M. Heinen

ABSTRACT

Despite having great potential for Hawaii and areas with similar climates, commercial prawn culture has not lived up to expectations, principally because appropriate technology for economical pond culture has not been adopted by farmers. During the last decade, results from research and development efforts initiated in Hawaii, and subsequently continued in Israel and Mississippi, document the importance of efficient pond management of heterogeneous individual growth (HIG) in achieving commercial success in both tropical and temperate environments.

Proof-of-concept trials in Hawaii using innovative stock manipulation which emphasized size-grading of nursed juveniles achieved increased production that was attributed to not only a larger starting size of the size-graded classes, as expected, but also a significant increase in the proportion of harvested animals composing the larger size classes. A follow-up field trial program, based upon the size-grading technology to manage and effectively reduce prawn HIG, was expanded in Mississippi to include larger experimental ponds and non-graded control groups, size matched to graded groups. Results of the most recent (1988) field trial and a corresponding economic analysis indicate that such a procedure is an economically valuable management tool that will contribute to the realization of successful commercial culture of prawns in the United States.

*Mississippi Agricultural and Forestry Experiment Station (MAFES) Publication.

INTRODUCTION

In Hawaii and other tropical regions, culture of prawns in earthen ponds is generally based upon a continuous cycle of stocking new postlarvae or small juveniles followed by a growout period and cull harvest of market-size individuals by seining (Malecha 1983). As discussed by Malecha (1983, 1988), production in Hawaii and climatically similar areas has characteristically been far below levels suggested by the early investigations of Fujimura and Okamoto (1972), despite a year-long growing season. Historically, development of the prawn aquaculture industry began well ahead of research. Growth of the industry suffered because an understanding of the biological characteristics of the prawn and their possible influence on limiting production were unknown. Subsequently, important biological and behavioral data relevant to the success of prawn farming were eventually obtained from laboratory and field trials. However, the survivors of an industry decline in Hawaii did not incorporate this critical information into their management strategies. Following Hawaii's "lead," industry in other areas also failed to embrace the new information and its representative technology.

Malecha (1988) recently outlined the reasons behind this disappointing outcome associated with commercial prawn aquaculture during the past decade. The traditional management system (Malecha 1983) currently practiced throughout the world was instituted with minimal knowledge of prawn biology and husbandry. Production of other agricultural species is usually limited by a function of energy-based inputs (feeds, fertilizer) and animal biomass (density and size). In contrast, the major limiting factor in prawn production is related to social structure within populations.

Prawn populations characteristically exhibit heterogeneous individual growth (HIG), resulting in highly skewed (to the right) size distributions (Smith et al. 1978; Malecha et al. 1981a; Ra'anan and Cohen 1985; Sandifer and Smith 1985). This population phenomenon is basically confined to males and results in a wide range of size that is based upon differential growth caused by complex social interaction. Results of extensive studies conducted in Hawaii on both experimental and pilot scales demonstrated that a prawn's age and size are no indication of its growth capability; a runted prawn can undergo considerable "compensatory" growth when size-graded from a mature population into one with prawns of similar size (Li et al.; Malecha et al., a-f; Zacarias et al., Appendix A). Indeed, HIG has been shown to be primarily nongenetic in prawns (Malecha et al. 1984). Small-sized prawns retain an inherent capacity for compensatory growth in the absence of large males. Therefore, prawn size is largely an ephemeral condition brought on by nongenetic, intra-populational behavioral factors. Small males can, therefore, develop into subdominant males in ponds, and HIG is best managed in ponds by removal of all large dominant males because their presence restricts the size transformation of subdominant and small males into dominant and subdominant males, respectively.

Malecha (1983, 1986) hypothesized that the traditional practice of periodic stocking and cull harvest by seine is inefficient because not all market size individuals are removed. Additionally, populations of smaller individuals persist, and sequentially stocked cohorts do not differentiate into distinct modal classes that can be separately harvested. Under current management practices in Hawaii, production and corresponding revenue from a pond cannot possibly achieve their full potential. Therefore, management practices designed to effectively control the detrimental effects of the social structure within a pond needed to be evaluated.

To reverse the trend of diminishing pond production in Hawaii and other areas, Malecha et al. (1981b) proposed an alternative management system. This system emphasized stock manipulation, pre-harvest size grading, and efficient harvest by pond draindown whereby market-size animals were removed and undersized individuals were restocked. Malecha (1983) described a commercial level system which used size grading and was instituted by a prawn producing company in Hawaii. The potential benefit of such grading procedures was recently validated during four years of trials conducted in experimental ponds in Hawaii (Malecha et al. a, Appendix A). Control and size-graded groups were similarly stocked and harvested. During the middle of the growout period, prawns in the size-graded groups were separated into two or three size classes and restocked. As a result, small individuals were afforded an opportunity to undergo compensatory growth in the absence of larger individuals. When compared to the traditional stocking-cull-harvesting-stocking system, this size grading approach was estimated to achieve potentially a 19% increase in production within a shorter period of growing (production) time.

Increased pond production through management of social structure has also been demonstrated in temperate climates through the technique of selective harvesting of pond populations (Willis and Berrigan 1978; Cohen and Ra'anan 1983). "Market size" prawns are selectively removed by seining a few weeks prior to total harvest by draining that is conducted before the advent of lethally low water temperatures. This selective seine and cull procedure succeeds in promoting the growth and production of the unculled (smaller) individuals. However, Ra'anan and Cohen (1983) observed that approximately 25% of prawns had not achieved a marketable size when final harvest occurred in temperate zone ponds in Israel. In an attempt to overcome this problem, Israeli investigators, one of whom participated in the initial field trials in ponds in Hawaii (Malecha 1984), evaluated stocking of larger juvenile prawns obtained from a grading procedure; mean weight, overall yield, and the percentage of marketable prawns at harvest improved. Karplus et al. (1986) evaluated the effect of separately stocking juveniles, graded into two size groups consisting of the upper one third and the lower two thirds numerically, on population structure and production. At harvest, after a growout period of 105 days, the percentage of small males in the upper-graded treatment (3%) was considerably lower than that of the control treatment (37%), resulting in a 25% increase in mean weight and gross yield. Another study (Karplus et al. 1987) evaluated the effect of size grading into three fractions, upper (32%), middle (45%) and lower (23%). Net income realized from the upper fraction was almost nine times that from the lower fraction and was attributed to differences in the population structure of males.

In Mississippi a well established and thriving catfish aquaculture industry led to research into the potential of alternative or supplemental aquaculture species. Initial investigations concentrated on seasonal monoculture of freshwater prawns in earthen ponds and the influence of both stocking density and stocking weight on mean harvest weight and yield. A positive relationship between yield and stocking density and an inverse relationship between stocking density and mean harvest weight were observed (D'Abramo et al. 1989). Even with a total growing season of approximately five months, the high incidence of small males in harvested populations significantly restricted revenue potential. Economic analysis of the production data revealed that the probability of consistently achieving any profit was low if management procedures were not modified (Clardy et al. 1985; Fuller et al. 1988). Therefore, a management practice was needed to minimize the detrimental effects of social structure (i.e., a large proportion of small males and correspondingly lower yields), and a size grading approach similar to that developed in Hawaii and tested in Israel appeared promising. Consequently, a field trial program was jointly designed and sponsored by the Hawaii Aquaculture Company and the Mississippi Agricultural and Forestry Experiment Station. The program is based upon the technology developed in Hawaii that is being transferred to Mississippi for testing. Overall, the goal of the size-grading approach in Mississippi is to increase overall mean harvest weight and production. Results of the 1988 field trial in Mississippi and a corresponding economic analysis are presented here.

MATERIALS AND METHODS

Postlarvae obtained from Blue Lobster Farms, Madera, CA and Sica Guadeloupe de Aquaculture, Pointe a Pitre, Guadeloupe were reared indoors in plastic lined swimming pools at the South Farm Aquaculture Unit, Mississippi Agricultural and Forestry Experiment Station, Mississippi State University. Subsequent to this nursery phase, juvenile prawns (mean wet weight = 0.22 g) collected from two pools were placed within a 5 mm square mesh nylon net enclosure supported by floats within a tank and allowed to passively grade themselves. Only a small proportion of the population, with a mean weight ±SE of 0.37 ± 0.01 g was retained within the net. The retained group was collected and moved to a temporary holding tank. Grading of the remaining prawns was attempted with a 4 mm mesh square nylon net, but the majority of prawns remained within the net. Manual size grading by visual inspection was then conducted. The resulting larger group was combined with the previously mesh-graded large group. The combined size-grading methods resulted in the numerical sorting of one-third of population into an upper graded portion and two-thirds into a lower graded portion. The upper-graded and lower-graded groups of juveniles, with mean ±SE wet weights of 0.30 ±0.01 and 0.14 ±0.01 g, served as two separate treatment groups. A third treatment group composed of ungraded juveniles (mean wet weight ±SE = 0.33 ±0.02) served as a control.

Pond Preparation and Stocking and Feeding of Prawns

Six earthen ponds located at the Coastal Aquaculture Unit of the Southern Branch of the Mississippi Agricultural and Forestry Experiment Station, Gulfport, MS were filled simultaneously with heated effluent from a power plant on 14-16 April 1988, approximately one month prior to stocking. Water surface area and mean water depth of each pond was 0.10 ha and 1.22 m, respectively. Crushed oyster shell was added at 1 ton/acre on 12 April and 13 May. Hydrated lime [$Ca(OH)_2$] was added periodically to each pond at 25 lbs/pond to raise pH levels. Each pond was fertilized with 2.7 kg of ammonium nitrate [$(NH_4)NO_3$] and 0.9 kg of phosphoric acid on 12 April and 13 May. On 12 April each pond also received 25 kg of cottonseed meal. Each pond was aerated daily with a 1 hp aerator from approximately 1600 to 0700 h. The graded and ungraded prawns were stocked on 13 May 1988 at a density of 29,652/ha, two ponds/treatment. Prawns were provided a custom made (Producers Feed Co., Isola, MS) 28% crude protein sinking pelleted feed (D'Abramo et al. 1989) twice daily (1/2 the daily ration/feeding) between 0700-0800 hr and between 1500-1600 hr, at weight dependent rates shown in Table 1. Ponds were sampled every three weeks to determine mean weights for adjustment of feeding rates.

Table 1. Weight dependent feeding rates for pond growout of *Macrobrachium rosenbergii.*

Mean wet weight (g)	Daily feeding rate* %
< 5	0
5-15	7
15-25	5
> 25	3

*(as-fed weight of feed/wet biomass of prawns x 100)

Water Quality

All ponds were checked twice daily at approximately 0600 and 1500 hr for dissolved oxygen levels and temperatures with a YSI model 57 meter. Salinity and pH were checked daily during mid-afternoon with a refractometer (Aquafauna Bio-Marine, Inc., Hawthorne, CA) and an Orion (model 301) meter, respectively. Total alkalinity was monitored weekly using a Hach (DREL 5) kit with a digital titrator. Decreasing total alkalinity (over 3 consecutive readings) was used as an indicator to predict potentially low (<7) pH problems.

Harvest

Ponds were harvested 25-28 October 1988; the total pond growout periods ranged from 165-168 days. Most of the prawns were removed by seining, and the remaining individuals were manually collected from the pond bottom subsequent to complete draining. Prawns harvested from each pond were bulk weighed and counted to determine mean weight, yield, and survival. Individual prawns were then separated into different morphotypes. Males were categorized as blue claw (BC), orange claw (OC), no claw (NC) or small male (SM) morphotypes. A male was classified as small if it weighed < 19 g. The ratio of the merus to the ischium segments of the first walking leg was used to separate BC males from OC males. To be classified as BC the ratio had to be greater than 1.0. BC or OC males that had lost both claws could not be positively identified and were designated no claw (NC) males. Females were separated into three morphotypes: ovigerous, open (previously ovigerous), and virgin (never having been ovigerous). Mean weights of all groups were calculated.

Statistical Analysis

Analysis of variance (ANOVA) for a completely randomized design and Fisher's Least Significant Difference (LSD) test (Steel and Torrie 1980) were used to determine significant ($P < 0.05$) differences among treatments relative to yield, mean harvest weight and percent composition of the morphotypes (SAS 1985). All percentages were arcsine transformed prior to the ANOVA test.

Economic Analysis

The synthetic firm or "economic engineering" approach was employed to evaluate the economic potential of grading juveniles prior to stocking. A synthetic Mississippi freshwater prawn farm described in Clardy et al. (1985) and updated in Fuller et al. (1988) was used for the analysis. The farm consisted of thirty-two 2.02 ha ponds with a total water surface area of 50.5 hectares. Total land requirements for the farm were 65.97 ha, which included 1.21 ha as a service area. Total investment requirements including land, buildings, and equipment were U.S. $631,541. Annual ownership costs for the farm were U.S. $72,431 and included depreciation, taxes, insurance, and interest on average investment. Some components of annual operating costs, specifically nursed juveniles, feed, harvesting-hauling and interest on operating capital, varied among production scenarios. These costs are described later. Costs of repairs and maintenance, fuel, chemicals, hired labor, and liability insurance did not vary and totaled U.S. $52,982, annually.

Based upon data obtained from the pond growout trials and the synthetic farm, three production scenarios were defined. Scenarios I, II, and III are based upon treatments consisting of upper-graded, lower-graded, and

Table 2. Ex-vessel prices of headless marine shrimp by count category for the northern Gulf of Mexico during September, October, and November, 1988. Source: National Marine Fisheries Service (1988).

Tail count (number/lb)	Price (U.S. $/lb)
< 16	7.75
16-20	6.76
21-25	5.74
26-30	4.64
31-35	3.75
36-40	3.15
41-50	2.64
51-60	2.51
> 60	2.38

ungraded juveniles, respectively. Revenues for each scenario were estimated using 1988 ex-vessel prices for headless marine shrimp harvested from the northern Gulf of Mexico during the months of September, October and November (Table 2). Use of 1988 prices for headless marine shrimp served a two-fold purpose. Analysis of historical trends in the price of marine shrimp revealed that prices for medium size tails have accordingly decreased as the volume of imports have increased. Therefore, the 1988 data probably provide the most accurate assessment of potential revenue. In addition, prices associated with marine shrimp were used because prices of freshwater shrimp were assumed to be unreliable. Such an approach is purposely conservative and does not imply that premium prices for large, whole prawns cannot be consistently obtained on a large scale. Fuller et al. (1986) presented a model consisting of log-log regression equations (Table 3) that related headless weight to whole weight for the aforementioned morphotypes of freshwater prawns. Using this model, tail weights were

Table 3. Values for regression equations used to determine tail weight from whole body weight of the different morphotypes in harvested populations of *M. rosenbergii*. Equations are expressed as log TW = a + b log WW, where TW = tail weight and WW = whole weight.

Morphotype*	a	b	R^2
BC	-0.23659	0.87854	0.89
OC	0.05637	0.75304	0.95
SM	-0.20985	0.94903	0.99
OV	-0.21078	0.97324	0.98
OP	-0.24250	0.96668	0.96
V	-0.11511	0.87530	0.97

* Morphotype designations are: BC = blue claw male, OC = orange claw male, SM = small male, OV = ovigerous female, OP =open (previously ovigerous) female, V = virgin (never having been ovigerous) female.

calculated by count category for each scenario and corresponding prices were used to estimate revenue for each scenario.

Cost of ungraded juveniles for stocking of ponds was estimated at U.S. $24/1,000 as suggested by the study of Leventos (1986). Grading costs of U.S. $10/1000 were exclusively assigned to the upper graded class of juveniles because this class was perceived as having greater value than the lower graded class. Cost of feed was based upon an estimate of U.S. $275.00 per metric ton. Harvesting-hauling costs were estimated to be U.S. $8.85 per 100 kilograms of shrimp harvested (Waits and Dillard 1987). An annual rate of 11% for six months for all items except feed and harvesting-hauling was assumed to calculate interest on operating capital. To estimate interest on operating capital for feed, an annual interest rate of 11 % for three months was used. No interest was calculated for harvesting-hauling.

RESULTS

All early morning oxygen readings were > 4.0 ppm and ranged from 4.0 to 11.8 ppm ($\bar{x} \pm SD = 7.3 \pm 0.75$). Afternoon oxygen readings ranged from 5 to 16 ppm ($\bar{x} \pm SD = 8.1 \pm 0.75$). No significant differences in either morning or afternoon dissolved oxygen readings existed between ponds for the growout period. Temperatures ranged from 15.2 to 31.0 C. Mean ±SD

Table 4. Survival, mean harvest weight, and yield of ponds stocked with graded and ungraded juvenile prawns. Survival means reported as back-transformed means of arcsine transformed values.

Treatment (Stocking weight, g)	Pond designation	Survival %	Mean weight g	Yield kg/ha	Feed conversion ratio[a]
Upper graded (0.30)	CAU 10	77.9	45.7	1,033	2.6
	CAU 8	89.7	44.1	1,144	2.1
	mean	84.3	44.9	1,089	2.4
Lower graded (0.14)	CAU 5	78.5	30.8	801	3.0
	CAU 7	74.4	37.3	859	3.2
	mean	76.5	34.1	830	3.1
Ungraded (0.33)	CAU 6	80.2	35.6	848	2.8
	CAU 9	73.1	29.8	648	3.5
	mean	76.7	32.7	748	3.2

[a] Ratio of total as-fed weight of feed fed to wct biomass produced

recorded pH was 7.6 ± 0.40 and ranged from 6.6 to 9.0. Alkalinity ranged from 6 to 68 ppm ($\bar{x} \pm SD = 24.7 \pm 3.8$). Salinity ranged from 0 to 10 ppt ($\bar{x} \pm SD = 3.2 \pm 1.7$).

Survival, mean wet weight, and total yield of each pond according to treatment are given in Table 4. Survival in all ponds ranged from 73.1 to 89.7% and no significant difference existed among treatments. Mean yields of the upper and lower graded treatments exceeded that of the ungraded control treatment by 45.6% and 11.0%, respectively. Mean wet weight of prawns harvested from the upper graded treatment was 37.3% and 31.7% greater than the ungraded and lower graded treatments, respectively. Mean harvest weight and yield of the upper graded treatment significantly exceeded ($P < 0.05$) that of the ungraded treatment.

Mean harvest weights of male and female morphotypes for each pond are provided in Table 5. The only significant difference ($P < 0.05$) between mean harvest weights of each morphotype was between small males in the upper graded and ungraded treatments. Differences in harvest mean weight of prawns and yield are apparently associated with changes in the relative

Table 5. Mean weights (g) of prawn morphotypes harvested from ponds stocked with graded or ungraded classes. BC = blue claw, OC = orange claw, SM = small, NC = no claw, OV = ovigerous, OP = open (previously ovigerous), V = virgin (never having been ovigerous).

		Males				Females		
Treatment	Pond	BC	OC	SM	NC	OV	OP	V
Upper graded	CAU 8	68.6	63.1	8.4	55.8	28.5	25.8	19.0
	CAU 10	63.3	68.4	9.5	49.4	26.8	34.3	24.7
	mean	66.0	65.8	9.0	52.6	27.7	30.1	21.9
Lower graded	CAU 5	60.4	55.9	8.1	38.8	25.5	24.0	17.9
	CAU 7	77.1	66.5	7.6	44.3	30.4	28.1	26.4
	mean	68.8	61.2	7.9	41.6	28.0	26.1	22.2
Ungraded	CAU 6	84.6	67.7	6.5	69.8	34.2	32.4	16.1
	CAU 9	64.7	62.8	5.4	46.6	30.1	27.7	15.6
	mean	74.7	65.3	6.0	58.2	32.2	30.1	15.9

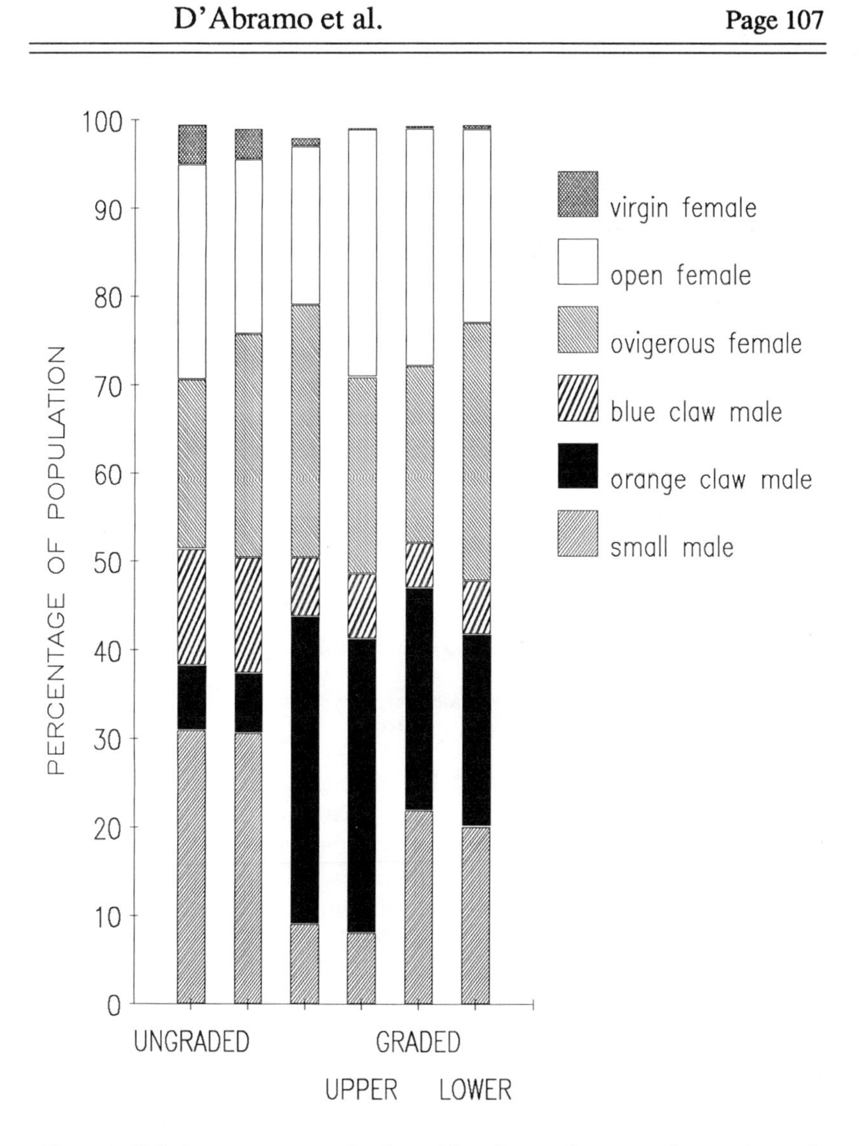

Figure 1. Relative percentages of male and female morphotypes at harvest in ponds stocked with upper-graded, lower graded, and ungraded juvenile prawns. Cumulative percentages that are < 100 are due to percentage of no claw male.

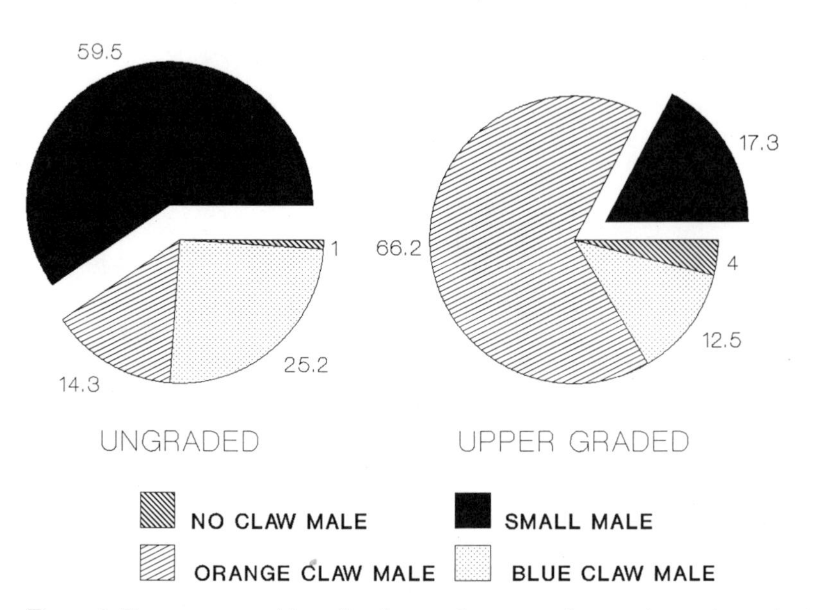

Figure 2. Percent composition of male morphotypes at harvest in ponds stocked with upper graded versus ungraded juveniles.

proportions of male morphotypes in the harvested population (Fig. 1). Pond populations of the upper-graded treatment averaged 8.6% small males, in contrast to 30.8% found in populations of the ungraded treatment. The percent compositions of male morphotypes in the pond populations of the upper-graded and ungraded treatments are shown in Figure 2. The upper-graded treatment had significantly smaller ($P < .001$) percentages of blue claw and small males than the ungraded treatment and significantly greater ($P < .001$) percentages of orange claw males. Accordingly, the percent contribution to the total harvested biomass made by orange claws greatly increased from 15.1 to 52.7%. All treatments differed significantly ($P < 0.05$) from one another in percentages of both small males and orange claw males. Percentages of blue claw males in the graded treatments were not significantly different, but both were significantly lower than that of the ungraded treatment. The percentage of virgin females of the ungraded treatment was significantly greater ($P < 0.05$) than that of either the upper or lower graded treatment. No significant treatment differences between the percentages of the remaining defined morphotypes were found.

Economic Analysis

The percentages of total revenue derived from each morphotype in each pond are presented in Table 6. Total revenue was additionally divided according to tail count category for each morphotype class (Appendix B). Estimated mean revenue of the two ponds representing each treatment was then extrapolated to derive gross returns for the synthesized 66 ha farm. Returns to land and management for the synthesized farm were U.S. $48,201 for the upper-graded juveniles (Scenario I), U.S. $25,514 for the lower-graded juveniles (Scenario II), and U.S. $-12,905 for the ungraded juveniles (Scenario III). The economic benefits of size-grading juveniles are clearly evident under the growing conditions described in this study. Even the lower-graded juveniles, which were stocked at less than half the weight of the ungraded ones, returned over U.S. $22,000 more to land and management than their ungraded counterparts.

Previous analyses have focused on exclusive use of one, rather than a combination, of the classes resulting from the grading procedure. In reality, a producer would desire the upper-graded juveniles, the class of individuals

Table 6. Percent of total revenue derived from each morphotype in populations of *M. rosenbergii* harvested from ponds stocked with graded and ungraded classes. See table 5 for explanation of morphotype symbols.

		Males				Females		
Treatment	Pond	BC	OC	SM	NC	OV	OP	V
Upper graded	CAU 8	10.9	53.8	0.7	2.2	17.3	15.1	0.0
	CAU 10	9.0	62.5	1.1	3.2	16.2	7.5	0.4
	mean	10.0	58.2	0.9	2.7	16.8	11.3	0.2
Lower graded	CAU 5	10.0	54.6	4.0	1.1	15.3	14.9	0.1
	CAU 7	11.5	48.3	2.5	0.8	23.5	13.2	0.3
	mean	10.8	51.5	3.3	1.0	19.4	14.1	0.2
Ungraded	CAU 6	30.5	19.7	3.4	1.4	22.9	20.7	1.4
	CAU 9	27.3	18.4	3.5	2.0	29.8	17.8	1.2
	mean	28.9	19.1	3.5	1.7	26.4	19.3	1.3

providing the highest net return. The present study suggests that the lower-graded juveniles have a value ranging from relatively low to zero. To analyze the impact of utilizing both upper and lower-graded juveniles on a given farm, an additional scenario was created. Scenario IV assumed that one-third of the ponds were stocked with upper-graded juveniles and two-thirds with lower-graded juveniles and included the 1 cent sorting charge (U.S. $10/1000) for the upper-graded individuals. Under Scenario IV, returns to land and management were U.S. $33,075.

DISCUSSION

The dramatic increase in mean weight and yield associated with the grading procedure strongly suggests that the management of the social structure of prawns represents a common solution to different problems that currently plague development of commercial prawn culture in Hawaii,

Table 7. Costs and returns (U. S. dollars) for three scenarios of freshwater prawn production, 66 ha farm, 1988.

Scenarios	I[1]	II[2]	III[2]	IV[3]
Gross returns	269,690	230,516	198,404	243,574
Annual ownership costs	72,431	72,431	72,431	72,431
Annual operating costs (constant for all scenarios)	52,982	52,982	52,982	52,982
Other annual operating costs				
Juveniles[1]	50,371	36,691	36,691	41,251
Feed	34,515	33,603	32,043	33,907
Harvesting/hauling	4,557	3,440	3,139	3,812
Interest on operating capital	6,634	5,856	5,772	6,115
Total annual costs	221,489	205,003	202,309	210,499
Returns to land and management	48,201	25,514	(12,905)	33,075

[1]Assumes a seed cost of $34/1,000.

[2]Assumes a seed cost of $24/1,000.

[3]Assumes one-third of ponds stocked with upper graded, two-thirds stocked with lower graded and a seed price of $34/1,000 for the upper graded and $24/1,000 for the lower graded juvenile prawns.

Mississippi and elsewhere. In particular, the percentage of small, low market value, males was reduced from 31.0 to 8.5 % of the pond population. This reduction was accompanied by a corresponding increase in orange claw males and 45.5% and 37.3% increases in mean harvest yield and mean harvest weight, respectively. Karplus et al. (1986) used size-grading techniques and achieved a 25% increase in mean weight and yield. Our greater increases may be principally due to a longer growout period, 166 versus 105 days.

The grading procedure effectively removes the social structure and corresponding skewed size distribution that has already been established in the nursery population. Large, fast growing juveniles are separated from smaller individuals that can be considered "root stock" needed for development of the rapid growers. Yet, the lower-graded treatments had greater yields than the ungraded treatment. These results are particularly encouraging because the stocking weight of the lower-graded prawns was 57.5% less than that of the ungraded ones. The grading procedure apparently succeeds in reducing weight variation by creating two new populations with normal distributions. Prawns in the lower-graded class respond to the removal of the larger individuals by increasing their growth rate (compensatory growth) and establishing a new social hierarchy. This response of lower-graded individuals warrants additional study.

The results clearly indicate that size grading of juvenile freshwater prawns is a superior production management alternative to stocking ungraded populations. However, the associated economic analyses presented are for comparative purposes and do not project that production of freshwater prawn is economically feasible, because the criteria for comparative evaluation were "returns to land and management." Individual entrepreneurs must decide whether the values depicted are sufficient to cover opportunity cost of land and necessary management returns, given the level of risk involved.

The data obtained in this study were applied to an analysis based upon a 66 ha farm in a Mississippi delta location. Variations in farm size and/or resource base often affect the per-unit costs of production. Therefore, studies are needed to evaluate net returns for various farm sizes, resources, and price situations. Rapid and reliable grading techniques need to be developed and tested, and estimates of associated costs must be included. Additional research should be directed toward evaluation of effects of the grading procedure when different percentages of upper and lower graded individuals and higher stocking densities are used. The grading procedure also needs further evaluation through inclusion of additional control groups in an experimental design similar to that used by Karplus et al. (1986, 1987). Significant additions to net returns may be realized from other areas that remain to be investigated. For example, if comparable results could be obtained in larger 8 ha ponds, potential cost savings would be approximately 9 percent, (Clardy et. al. 1985), an amount which implies an increase in returns to land and management of about $18,000.

Production scenarios presented in this study are derived from one year's data that were obtained from only two replicates for each treatment (six ponds). Although the results of this investigation are encouraging, they should be replicated and extended to verify commercial application. Nevertheless, these results have led to a major reassessment of the prospects of achieving economic feasibility of freshwater prawn farming in temperate and tropical regions of the world.

ACKNOWLEDGMENTS

We thank the manager, Mr. Mike Murphy, and staff, Ms. Patti Simms, Mr. Rick Kastner and Mr. Frank Backman, of the Coastal Aquaculture Unit, Southern Branch, Mississippi Agricultural and Forestry Experimental Station. Their able assistance and dedication to the daily management of the experimental ponds, as well as sampling, harvesting, and data collection were major factors in the successful completion of this investigation. Special thanks are extended to Dr. Fred Tyner, Director, Southern Branch, MAFES, for his ongoing interest and cooperation. Comments of two reviewers served to improve the content and presentation of the manuscript and are greatly appreciated.

LITERATURE CITED

Clardy, G. N., M. J. Fuller and J. E. Waldrop. 1985. Preliminary economic evaluation of freshwater shrimp production in Mississippi. Agricultural Economics Research Report 159, Mississippi State University, 46 pp.

Cohen, D. and Z. Ra'anan. 1983. The production of the freshwater prawn *Macrobrachium rosenbergii* (de Man) in Israel: Improved conditions for intensive monoculture. Bamidgeh 35:31-37.

D'Abramo, L. R., J. M. Heinen, H. R. Robinette, and J. S. Collins. 1989. Production of the freshwater prawn *Macrobrachium rosenbergii* stocked as juveniles at different densities in temperate zone ponds. Journal of the World Aquaculture Society 20:81-89.

Fuller, M. J., Y. N. Lin and J. E. Waldrop. 1986. A prediction technique to determine tail yields of *Macrobrachium rosenbergii*. Paper presented at Annual World Mariculture Society meeting, Reno, Nevada.

Fuller, M. J., D. W. Whitten and G. D. Thomas. 1988. Economic analysis of alternative stocking rates of freshwater shrimp, 1986. Agricultural Economics Research Report 180, Mississippi State University.

Fujimura, T. and Okamoto, M. 1972. Notes on progress made in developing a mass culturing technique for *Macrobrachium rosenbergii* in Hawaii. Pages 313-327 *in* T.V.R. Pillay.editor. Coastal Aquaculture in the Indo-Pacific Region. Fishing News Books, Ltd., London.

Karplus, I., G. Hulata G. W. Wohlfarth, and A. Halevy. 1986. The effect of size-grading juvenile *Macrobrachium rosenbergii* prior to stocking on their population structure and production in polyculture. I. Dividing the population into two fractions. Aquaculture 56:257-270.

Karplus, I., G. Hulata, G. W. Wohlfarth, and A. Halevy. 1987. The effect of size-grading juvenile *Macrobrachium rosengergii* prior to stocking on their population structure and production in polyculture II. Dividing the population into three fractions. Aquaculture 62:85-95.

Leventos, T. P. 1986. A preliminary economic analysis of hatching and nursing costs for freshwater shrimp seed stock. M.S. thesis, Department of Agricultural Economics, Mississippi State University. 67 pp.

Malecha, S. R. 1983. Commercial pond production of the freshwater prawn *Macrobrachium rosenbergii* in Hawaii. Pages 231-259 *in* J. P. McVey, editor. CRC Handbook of mariculture, Vol. I. Crustacean aquaculture. CRC Press, Inc., Boca Raton, FL.

Malecha, S. R. 1984. Development of a new management technology for commercial prawn farming, Phase I. Job completion report, grant No. OCE-8113995. National Science Foundation- SBIR program.

Malecha, S. 1986. New techniques for the assessment and optimal management of growth and standing crop variation in the cultured freshwater prawn, *Macrobrachium rosenbergii*. Aquacultural Engineering 5:183-197.

Malecha, S. R. 1988. Recent advances in the production of freshwater prawn. Proceedings of the Aquaculture International Congress, Vancouver, British Columbia, Canada: 583-591.

Malecha, S. R., D. Bigger, T. Brand, A. Levitt. S. Masuno, and G. Weber. 1981a. Genetic and environmental sources of growth pattern variation in the cultured freshwater prawn, *Macrobrachium rosenbergii*. Paper presented at the World Conference on Aquaculture Sept. 21-25, 1981, Venice, Italy.

Malecha, S. R., J. Polovina and R. Moav, 1981b. A multi-stage rotational stocking and harvesting system for year round culture of the freshwater prawn, *Macrobrachium rosenbergii*. University of Hawaii Sea Grant Technical Report, UNIHI-SEAGRANT-TR-81-01, 33 pp.

Malecha, S. R., S. Masuno and D. Onizuka. 1984. The feasibility of measuring the heritability of growth pattern variations in juvenile freshwater prawn, *Macrobrachium rosenbergii* (de Man). Aquaculture 38: 347-36.

National Marine Fisheries Service. 1988. Shrimp statistics. National Oceanographic and Atmospheric Administration, U.S. Department of Commerce, New Orleans, LA.

Ra'anan, Z. and D. Cohen. 1983. Production of the freshwater prawn *Macrobrachium rosenbergii* in Israel. II. Selective stocking of size subpopulations. Aquaculture 31:369-379.

Ra'anan, Z. and D. Cohen. 1985. The ontogeny of social structure in the freshwater prawn *Macrobrachium rosenbergii*. Pages 277-311 *in* A. Wenner and F. R. Schram, editors. Crustacean issues II: crustacean growth. A. A. Balkema Publishers, Rotterdam.

Sandifer, P. A. and T. I. J. Smith. 1985. Freshwater Prawns. Pages 63-125 *in* J. V. Huner and E. E. Brown, ediotrs. Crustacean and mollusk aquaculture in the United States. AVI Publishing Co., Inc., Westport, CT.

SAS Institute Incorporated. 1985. SAS/STAT Guide for personal computers, Version 6 Edition, Cary, N.C. 378 pp.

Smith, T. I. J., P. A. Sandifer, and M. H. Smith. 1978.Population structure of Malaysian prawns, *Macrobrachium rosenbergii* (de Man) reared in earthen ponds in South Carolina, 1974-1976. Proceedings of the World Mariculture Society 9: 21-38.

Steel, R. G. D., and J. H. Torrie. 1980 Principles and procedures of statistics. McGraw-Hill, Inc. New York, NewYork.

Waits, W. J., III and J. G. Dillard. 1987. Costs of processing and hauling freshwater shrimp in Mississippi. Information Bulletin 953. Mississippi Agricultural and Forestry Experimental Station., Mississippi State, MS. 22 pp.

Willis, S. A. and M. E. Berrigan. 1978. Effects of fertilization and selective harvest on pond culture of *M. rosenbergii* in central Florida. Completion Report for U.S.Department of Commerce, NOAA, NMFS., PL 88-309, No.2-298-R-a, Job 3B, Gainesville, FL.

APPENDIX A

Li, D., L. E. Barck, S. R. Malecha, S. Masuno, R. Stanley and D. Zacarias. Biology and management of growth variation in the freshwater prawn, *Macrobrachium rosenbergii* III. The effect of metabolic and behavioral factors on heterogenous individual growth (HIG). (manuscript in preparation).

Malecha, S. R., L. E. Barck, E. R. MacMichael, S. Masuno, and D.Zacarias. a. Biology and management of growth variation in the freshwater prawn, *Macrobrachium rosenbergii* VII. The effect of pre-harvest size grading and harvest efficiency in pond production. (manuscript in preparation).

Malecha, S. R., L. E. Barck, E. R. MacMichael, S. Masuno, G.Webber, and D. Zacarias. b. Biology and management of growth variation in the freshwater prawn, *Macrobrachium rosenbergii* II. Compensatory growth potential of individual runted juveniles. (manuscript in preparation).

Malecha, S. R., L. E. Barck, E. R. MacMichael, S. Masuno, D. Zacarias. c. Biology and management of growth variation in the freshwater prawn, M*acrobrachium rosenbergii* IV. The effect of individual and communal rearing conditions on heterogenous individual growth (HIG). (manuscript in preparation).

Malecha, S. R., L. E. Barck, E. R. MacMichael, S. Masuno, and D. Zacarias. d. Biology and management of growth variation in the freshwater prawn, *Macrobrachium rosenbergii* V. Evidence for a water borne factor determining heterogenous individual growth. (manuscript in preparation).

Malecha, S. R., S. Masuno, and D. Zacarias. e. Biology and management of growth variation in the freshwater prawn, *Macrobrachium rosenbergii* VI. Compensatory growth in sized, graded laboratory populations. (manuscript in preparation).

Malecha, S. R., P. A. Nevin, P. Ha-Tamura, L. E. Barck, Y. Lamadrid-Rose, S. Masuno, and D. Hedgecock. f. Production of progeny from crosses of surgically sex-reversed prawns, *Macrobrachium rosenbergii*: duplications for commercial culture. Aquaculture (in press).

Zacarias, D., S. R. Malecha, L. E. Barck, E. R. MacMichael, S. Masuno, P. Ha-Tamaru, and D. Hedgecock. Biology and management of growth variation in the freshwater prawn, *Macrobrachium rosenbergii* I. Relationship between larval, juvenile and adult growth using allozymes as genetic tags. (manuscript in preparation).

APPENDIX B

Total whole weight of each morphotype according to count category, total weight and revenue for each count category, and total revenue of freshwater shrimp harvested from experimental ponds. Explanation of morphotype symbols is provided in Table 5.

CAU 8:

Count category	Total weight (kg) Male BC	OC	SM	NC	Female OV	OP	V	Total weight/ category (kg)	Total revenue/ category ($U.S.)
< 16	0.814	4.715		0.289				5.189	99.44
16-20	1.467	8.727		0.346				10.540	157.11
21-25	0.973	3.653		0.060	1.443			6.129	77.57
26-30	0.542	0.914		0.034	4.728	2.250		8.468	86.63
31-35	0.216	0.762		0.014	2.270	3.603		6.865	56.77
36-40	0.036				0.342	2.892		3.271	22.73
41-50		0.295			0.105	0.754	0.010	1.163	6.77
57-60		0.114	0.026					0.140	0.78
> 60		0.093	0.699					0.791	4.15
TOTAL									511.95

CAU 10:

Count category	Total weight (kg)							Total weight/ category (kg)	Total revenue/ category ($U.S.)
	Male				Female				
	BC	OC	SM	NC	OV	OP	V		
< 16	0.969	11.219		0.539				12.727	217.50
16-20	1.253	7.266		0.360				8.879	132.39
21-25	0.658	1.346		0.082	0.387	0.126		2.599	32.90
26-30	0.199	0.664		0.035	4.433	0.740	0.065	6.135	62.76
31-35	0.105	0.186		0.057	3.621	1.179	0.083	5.231	43.26
36-40	0.063	0.473		0.085	0.741	2.459	0.062	3.883	26.99
41-50	0.102		0.065	0.011		0.486	0.059	0.723	4.21
51-60			0.168			0.058		0.226	1.25
> 60			0.817				0.005	0.822	4.31
TOTAL									525.57

CAU 5:

Count category	Total weight (kg) Male BC	OC	SM	NC	Female OV	OP	V	Total weight/ category (kg)	Total revenue/ category ($U.S.)
< 16	0.530	3.842		0.072				4.444	75.95
16-20	1.417	7.791		0.112				9.374	139.77
21-25	0.335	2.275		0.020	0.303			2.933	37.13
26-30	0.308	0.323		0.051	3.051	0.501		4.233	43.30
31-35	0.087	0.395		0.069	2.084	3.239	0.013	5.888	48.69
36-40	0.146	0.242		0.024	0.966	2.804	0.012	4.192	29.13
41-50	0.019		0.736	0.020	0.322	1.295	0.041	2.432	14.15
51-60			0.359			0.093		0.452	2.50
> 60			1.885		0.054		0.007	1.947	10.20
TOTAL									400.82

CAU 7:

Count category	Total weight (kg)							Total weight/ category (kg)	Total revenue/ category ($U.S.)
	Male				Female				
	BC	OC	SM	NC	OV	OP	V		
< 16	1.742	10.520		0.117				12.379	211.56
16-20	1.292	3.515		0.055	0.248			5.109	76.18
21-25	0.467	0.625		0.019	3.005	0.944		5.060	64.06
26-30	0.259	0.125		0.048	6.271	2.581	0.034	9.317	95.31
31-35		0.110		0.023	1.490	2.067	0.072	3.762	31.11
36-40				0.036		1.468	0.025	1.529	10.63
41-50			0.290			0.076		0.366	2.13
51-60			0.365				0.008	0.373	2.06
> 60			1.711				0.007	1.718	9.00
TOTAL									502.04

CAU 6:

Count category	Total weight (kg) Male				Female			Total weight/ category (kg)	Total revenue/ category ($U.S.)
	BC	OC	SM	NC	OV	OP	V		
< 16	4.828	4.458		0.337				9.624	164.47
16-20	2.871	0.720		0.051	2.119	0.241		6.002	89.49
21-25	0.874	0.177			3.591	2.970		7.612	96.37
26-30	0.428	0.097		0.016	2.085	3.627	0.097	6.350	64.96
31-35		0.082			0.753	1.253	0.134	2.222	18.38
36-40	0.062	0.074			0.092	0.599	0.209	1.036	7.20
41-50					0.162	0.487	0.240	0.889	5.17
51-60			0.217			0.087	0.114	0.417	2.31
> 60			2.749				0.146	2.895	15.17
TOTAL									463.52

CAU 9:

Count category	Total weight (kg)							Total weight/ category (kg)	Total revenue/ category ($U.S.)
	Male				Female				
	BC	OC	SM	NC	OV	OP	V		
< 16	1.700	1.426		0.180				3.307	56.52
16-20	1.713	1.773		0.131				3.617	53.93
21-25	1.711	0.449		0.081	2.130	0.730		5.102	64.59
26-30	0.877	0.171			5.849	2.432	0.049	9.377	95.93
31-35	0.265	0.109		0.030	0.755	1.687	0.043	2.888	23.88
36-40	0.059			0.037	0.446	0.904	0.049	1.495	10.39
41-50	0.053			0.011		0.532	0.280	0.876	5.10
51-60							0.051	0.051	0.28
> 60			2.130				0.122	2.252	11.80
TOTAL									322.42

OPERATIONAL PLANNING FOR THE SEMI-INTENSIFICATION OF AN EXTENSIVE MARINE SHRIMP FARM IN ECUADOR

Spencer Malecha, Lorena E. Barck, Elizabeth R. MacMichael, Thomas S. Desmond, Gil Kohnke, and Jonathan Roberts.

ABSTRACT

This paper discusses planning for the semi-intensification of an extensive marine shrimp farm in Ecuador with an emphasis on the following points: the importance of the variation as well as the mean production values in commercial hatcheries and ponds; production planning and the integration of hatchery and pond stocking phases; the calculation of critical standing crop values for a minimum level of feed application to large ponds; financial and cash flow planning using computer-based spreadsheet models; and practical considerations for growth management. The purpose of this paper is to illustrate some simple, direct approaches and concepts that were used in the strategic planning of the project. We do not generalize to other projects, but emphasize one particular project as a case study. We emphasize practical applications and planning rather than research or management.

INTRODUCTION

Ecuador has an established industry of extensive marine shrimp (*Penaeus vannamei*) culture which is based on the use of large growout ponds and wild-caught post larvae (PLs). Current shrimp production averages 300-500 kg/ha/yr for extensive production systems based on 2-3 growout cycles/year. In most cases the ponds are very large lake-like impoundments. As an example, pond sizes on the farm described here range from 6.0 hectares to 27.8 hectares. The PLs are either introduced directly into these ponds by tidal fluctuation and/or pumping, or are purchased from dealers who collect the PLs off-shore and transport them to the shrimp farms.

Until recently, the rate of return on the investment involved in developing a growout farm was extremely favorable. Indeed, land and labor and earth-work costs were low and there were almost no pond management costs beyond improving water quality. This was accomplished by periodically pumping water into ponds which were stocked at the low, extensive levels of approximately 2-3 PLs/m^2.

Until 1984, the availability of native PLs was sufficient to stock the existing ponds in Ecuador. Unfavorable meteorological conditions in 1984 however, caused the native seed to be in critically short supply. This situation stimulated the building of numerous hatcheries to compensate for the short-fall in native seed availability. Once a company owning only growout ponds makes a financial commitment to building a hatchery, it is necessary to cover the costs of that investment by increasing the output of the growout farm. Moreover, the price of raw materials, labor and land for farms has risen. All of these conditions have mandated semi-intensification of pond production throughout Ecuador.

METHODS

Planning for Production Semi-Intensification: Goals

With the intensification of production on the farm described here, we calculated that production could be increased from 300 kg/ha/yr to between 600 to 1,000 kg/ha/yr.

We set two major goals for our management strategy in order to achieve this level of production:

Goal 1. Design a hatchery physical plant and management program which would minimize the effects of fluctuating availability of wild Pls, at the same time cover all of its own costs, and supply the farm ponds with sufficient seed to stock the farm to capacity.

Goal 2. Design and implement a pond stocking and management strategy to recover the costs of the hatchery reared PLs, while at the same time increasing pond production to an economically optimal level within the confines of the existing physical limitations of the farm.

Approach

Our approach to achieving our goals used the following guidelines.

1. Revenues ("economic efficiency") are not necessarily maximized by biological efficiency since the variation in hatchery production is as important as mean production and can be planned for using the conceptual framework of continual process control monitoring (Jessup 1985).

In research, biological yield (e.g., kg/ha) is emphasized. In business, however, an economic return (after-tax revenue) is the ultimate goal and is not always accomplished by an increase in total project size or production. Our analyses showed favorable economic returns if we minimized initial hatchery capitalization and accommodated production variation with increased operating funds as opposed to high fixed costs using capital funds. In this regard, we increased the input of primary materials to the system (nauplii and gravid females) to achieve the desired output.

In designing the hatchery physical plant and management system we planned for a modest average biological production and took into account the existing variation in PL production. We did not attempt to reduce this variation but monitored it using the conceptual framework of "continual process control monitoring" (Jessup 1985) in order to determine whether it was within acceptable (i.e., economically tolerable) limits. In this way it was possible to design and operate a "minimal" facility to meet farm PL requirements and, at the same time, cover costs at a break-even level.

2. The front-end costs in shrimp culture production systems are relatively small when considered in the context of the increase in value of the "seed" as it grows into harvestable shrimp.

In shrimp culture, revenues are due to pond production. Therefore, costs of production of post larvae are much less important than PL availability. Indeed, the front-ended (i.e., hatchery) costs are relatively small compared to the revenue generated by the end product. Therefore, the most relevant consideration in our strategic planning was to insure that there would be enough PLs to stock the ponds on time, not whether hatchery survival (%) and production (PLs/l) were very high. Of course, survival and production are important, but beyond a minimally acceptable level, determined by us to be 40-50% survival and 40-50 PL/l (which is modest by industry standards), it does not pay to invest capital in order to drive survival and production higher. Indeed, the increase in value of PLs, when compared to their value as marketable shrimp at harvest, is at least an order of magnitude greater than their value at the end of the hatchery cycle.

Consider the following example. The selling price of PLs in Ecuador was $3-4/1,000 PLs. At a harvest size of 28 animals/lb (16 g/animal), the value of each harvested shrimp is 14 cents if the selling price is $3.90/lb. The increase in value of PLs as harvested shrimp is at least 35 times their value as PLs. Even if there is a 50% mortality in the grow-out pond, the increase in value of a PL when it is a harvestable shrimp is still 17-fold. This large increase in value can offset an increase in the front-ended costs of producing the PLs. Any reasonable increase in operating costs — even those that double or triple PL costs — can probably be absorbed by pond-generated revenues.

3. The emphasis should be placed on keeping the ponds stocked at all times since it is the ponds, not the hatchery, which are the income generators.

To keep the grow-out ponds stocked, we made use of nursery ponds as an intermediate step in the hatchery-to-growout pond transfer of post-larvae. This allowed the hatchery to produce on a calendar monthly schedule unencumbered by pond readiness. This also allowed the pond harvests to be scheduled to take advantage of current market price, weather, the available work force, and other factors. We attempted to correlate the availability of PLs from each hatchery cycle with pond availability using simple Gantt charting methods which consisted of visualizing how each hatchery production cycle could be stocked in a particular pond. Each pond stocking, growout period and harvesting schedule could therefore, also be visualized. This time-line charting of all production systems was tied into a computer model of financial analysis and profit sensitivity (point 5 below).

4. Feed costs can be minimized by applying simple principles of ecology to shrimp growth.

PLs stocked at the rate of 3 PLs/m^2 were not provided with supplemental feeding, so all growth and production was from the naturally-occurring pond food organisms. Therefore, the range of extensive production of 300-500 kg/ha/yr was used as an estimate of the range of critical standing crop (CSC) for the pond ecosystem. Critical Standing Crop (CSC) is defined by Hepher (1978) to be that instantaneous biomass level (density x mean size) in a pond at which growth rate is reduced from an optimum level for the particular conditions of the pond at the time. Supplemental feeding should begin when a population reaches its CSC since food is limiting and growth rate diminishes. In practical terms, CSC cannot be — nor does it need to be — estimated precisely since it is an instantaneous value. However, using simple graphic means and plotting pond sampling data on semi-log graph paper, it can be determined whether or not the population has exceeded its CSC and therefore should be fed or have the feed increased.

5. It is important to forecast the sensitivity of revenue to production variables as soon as possible.

As a commercial venture, it was important to determine how sensitive our revenues would be to several major production parameters such as stocking density and final harvest size which are subject to manipulation. Other parameters are directly related to fixed variable costs (food conversion efficiency and survival) and are not subject to direct manipulation. Theoretical and practical models determined for other shrimp operations (Griffin et al. 1984) did not give us the appropriate information for determining our revenues. We needed to know the sensitivity of our revenues to these parameters using the specific stocking and harvest schedules and our specific costs.

6. When quantitative information is lacking or inappropriate for a project, useful models can be built using the conceptual framework and heuristic value of the Delphi technique.

The Delphi method (Linstone and Turoff 1975) is a method of forecasting using expert opinion. When used formally this method gives an estimation of a quantitative value using a group of expert opinion. A detailed description of this method is not appropriate here and can be found in the literature along with its statistically demonstrated validity (Linstone and Turoff 1975; Zuboy 1981; Fresfeld and Foster 1971). With the Delphi approach to decision-making, a specific value can be estimated by using the opinion of experts who say what they believe the value to be based on their experience. We used the conceptual framework of the Delphi approach to estimate the values we needed to plan our feed management strategies and to develop the production functions for the financial model. In our case, to use an example, the length of time of the growth period of a pond population during a particular season was needed in order to estimate cash flow. As there were no published or measured quantitative information available to us nor, of course, had we the time or the resources to conduct years of growth experiments, we estimated the parameters we needed using the Delphi approach.

DISCUSSION

Hatchery Capital costs

A high capitalization cost for the hatchery would lead to high fixed costs. Rather than increasing capitalization in the hatchery, the most appropriate strategy for our project was to incur slightly higher fixed variable operational costs. These are costs associated with buying more sourced animals as well as larval feed in order to insure that pond stocking goals are met. Table 1 shows that the capital investment/liter capacity of our hatchery compares favorably to similar investments for other hatcheries which are reported in the literature. The first year's operating expenses are included in the $3.75/liter unit cost investment for our hatchery.

Hatchery Production Variation

In the hatchery, there are several dependent variables or process control steps (sensu Jessup 1985): number of eggs spawned/female, per unit time

Table 1. Comparative hatchery capital costs for our project and others reported in the literature.

	Species	Type	Capacity x1000 l	Capital Investment $US/l[1]	Reference
Ecuador	PV	MSH	64	$3.75[2]	This project
Honduras	PV, PS	MSH	96	15.62	Mock 1980
Ecuador	PV, PS	MSH	45	7.78	Mock 1981
Texas	PV, PS	H	60	7.60	Johns et al. 1981
Costa Rica	PV. PS	MSH	30	6.75	Pers. Comm.
Philippines	PM	H	24	0.16	SEAFDEC 1984

PV, PS, PM = *Penaeus vannamei*, *P. stylirostris*, *P. monodon*
M, S, H = Maturation, Spawning, Hatchery

[1] = total working volume capacity of larvae rearing tanks
[2] = also includes first year operating costs.

egg to nauplii survival, nauplii to zoea survival, zoea to mysis stage survival, mysis to postlarva survival and so on. All variation in the hatchery production system can either be described as due to "common" causes, which are sources of variation within a process that are in statistical control, or "special" causes which are all other sources of variation within a process which cannot be adequately explained by any single distribution of the process output as if the process were in statistical control (Jessup 1985).

Table 2 shows the average larval survival and production during the first ten cycles of the first year of production, October 1986 through October 1987. Larval survival is expressed as the number of post larvae recovered at the end of the cycle as a percentage of nauplii larvae stocked at the beginning of the cycle. Postlarvae production displayed no specific trends. However, survival tended to increase as better management strategies were worked out to meet the specific conditions of the hatchery. Despite the design and implementation of a complete maturation facility, the latter was not utilized since gravid females or nauplii were available to fulfill hatchery stocking requirements. Nauplii became available at very favorable prices as

Table 2. Hatchery production and survival for each cycle during the first year, start-up phase October 1986-October 1987. Mean weight of gravid females was 50 g.

Cycle	Production PL/l	Larvae survival %	Seed source	Amount purchased
1	20	14	nauplii[1]	1.8 million
2	21	37	gravid[2]	11 females
3	34	47	gravid	7 females
4	36	60	gravid	28 females
5	91	56	gravid	10 females
6	50	50	nauplii	3.2 million
7	33	40	nauplii	3.3 million
8	48	33	nauplii	4.56 million
9	0	0[3]	nauplii	3.9 million
10	36	38	gravid	59 females

[1] purchased nauplii
[2] purchased gravid females
[3] *Vibrio* disease.

larger hatcheries overproduced and released them to the spot market. Figure 1 shows the variation that is present in the per female nauplii output. A line connects the mean values, but the graph shows that the number of nauplii varies considerably per female. It was not possible to count the number of eggs spawned per female under the production conditions that existed. Females were purchased in a nearby coastal village from dealers who obtained them from collectors. In cycles 2, 3, 4 and 10 some females produced no nauplii, while some females (presumably from the same area as the other females and obtained, in many cases, from the same dealers) produced nauplii in the range of 10,000 nauplii per female to as high as 580,000 nauplii per female. Cycle 5 represented the largest overall mean variation of nauplii production; from 10,000 - 200,000 per female. Variation such as this must be taken into account when designing a hatchery management program.

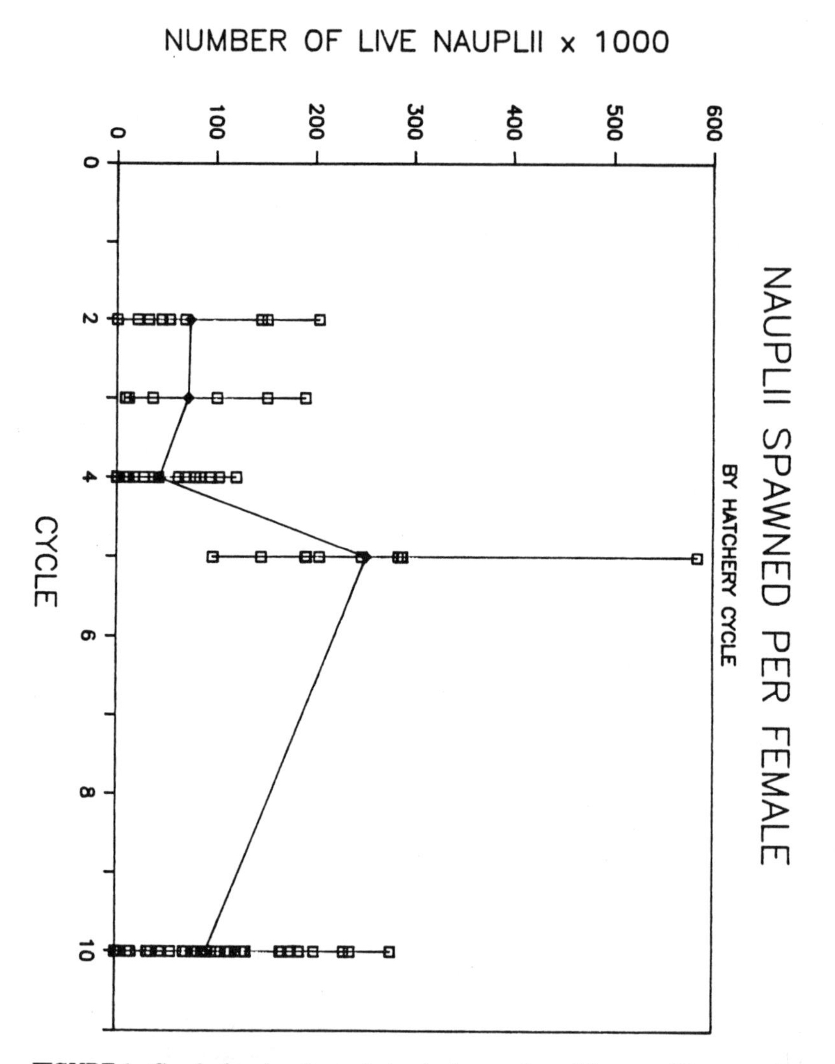

FIGURE 1. Graph showing the variation in the number of live nauplii spawned per gravid female purchased, in hatchery cycles 2 through 5 and 10. Values for each cycle are connected by lines to show cycle-to-cycle variation. Each data point represents values that could be assigned to each female which averaged approximately 50g each. The number of females purchased and spawned for each cycle is given in Table 1.

Each month between 7 and 59 females were purchased, depending upon the cycle and the PL requirements. In October of 1987, the price of each gravid female was approximately 3,500 sucres (Ecuadorian currency) or about $14.00. Purchasing additional females, despite the apparent high unit cost, had an insignificant impact on the overall operating expenses of not only the hatchery but the hatchery and grow-out farm considered as a single production system.

In cycles 1 and 6 through 9, nauplii were available on the spot market. These were obtained either from larger hatcheries who overproduced or from "hatching stations" on the north coast of Ecuador which obtained gravid females from northern shrimp populations. Figure 2 shows the percent survival to first PL of the purchased nauplii and those nauplii hatched from eggs spawned from purchased females. These estimates were obtained using volumetric methods. The percent survival to first PL is the total number of animals (zoea, mysids, and PLs) present in a larvae rearing tank on the morning that the first post larva was observed, and is expressed as a percentage of nauplii stocked in that tank. Each data point in Figure 2 represents a value for each tank, and they show a considerable amount of tank-to-tank variation in survival. In some cases data are not available due to logistic reasons (cycles 3 & 5) or are available for only 1 tank (cycles 2 & 4). Cycles 6 through 10 (Fig. 2) represent complete data sets. Cycle 9 represents complete mortality due to a *Vibrio* disease outbreak.

Figure 3 represents the survival of the nauplii to PL 30 and, as in Figure 2, reflects a considerable variation in tank-to-tank performance. A data point is not represented in cycle 9. Since all tanks are similarly constructed and managed, it was concluded that the variation shown in Figures 2 and 3 represents the normal "common caused" (Jessup 1985) variation and therefore could not be eliminated. We concluded that the basic unit of hatchery production — despite the conventional expression — is not one of volume (e.g., liter or ml) but the production per tank. The implications of this for strategic planning of hatchery production management were two-fold, assuming that the variance in mean tank performance could not be reduced. First, increasing the inputs to the system would increase total production despite a modest production per unit. Second, adding additional numbers of smaller tanks could reduce the risk of underproduction in a

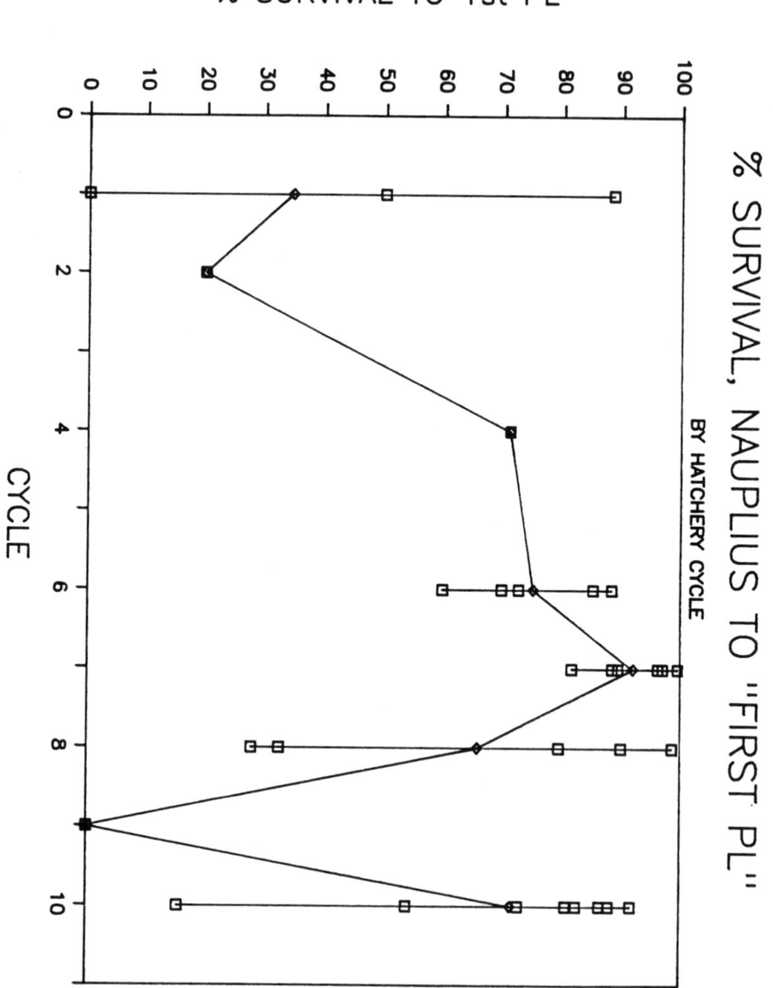

FIGURE 2. Variation in survival of nauplii expressed as the estimated number of animals in a larval rearing tank (zoea, mysis, PLs) on the day the first postlarva was observed as a percentage of the nauplii stocked into the tank on the first production day. Data were unavailable for cycles 3 and 5. Some cycles are represented by data from only one tank although more were stocked. Cycle 9 showed complete mortality due to a disease infection. Each data set is connected by a line at its mean value for each cycle.

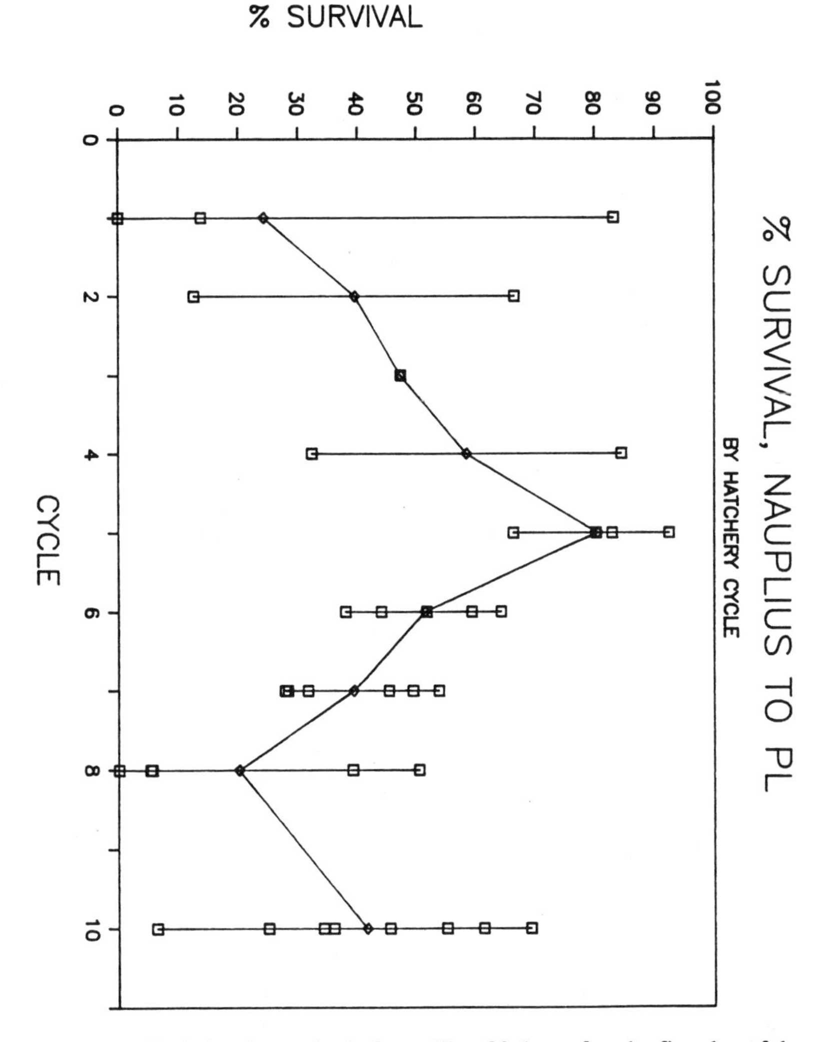

FIGURE 3. Variation in survival of nauplii to 30 days after the first day of the production cycle defined as when nauplii are stocked into the larval rearing tanks. Each data point represents one larval tank (some cycles have a smaller number of tanks represented than were used in the cycle). The survival of nauplii is represented as the estimated number of postlarvae harvested from the tank on "day 30" expressed as a percentage of nauplii stocked into the tank. Each data set is connected by a line at its mean value for each cycle.

particular cycle. Expansion of the hatchery and replacement of the hatchery tanks was, therefore, planned accordingly. Because of the wide variability in tank-to-tank performance, using a smaller number of larger tanks was not planned as an option. On the contrary, the planned hatchery expansion accounted for tanks of equal or smaller volume than those in use, thereby reducing the risk of underproduction.

Growout Pond Strategy

The strategic planning for the growout ponds had five major parts.

1. A shift from the use of native seed for stocking the ponds to the use of hatchery-reared PLs.

The use of wild-caught, native stocking cohorts involves mixed species with no control of the percentage of *Penaeus vannamei*. Suppliers provide the farms with native postlarvae which consisted of up to three other *Penaeus* species (*P. stylirostrus*, *P. occidentalis* and *P. californensis*). These species do not do well in ponds and a multi-species stocking results in a virtual mono-culture of *P. vannamei* at harvest.

2. An increase in the stocking density to semi-intensive levels and the implementation of a program of supplemental feeding when populations exceed the critical standing crop.

A priori, we estimated that stocking density could be increased to between 6-10 PL/m^2 based on densities reported by others for semi-intensive *P. vannamei* production using supplemental feeding (Lawrence et al. 1985).

The method we used to determine the feeding rates is described as follows. Growth, expressed as weight (W) at time, t can be expressed as:

$$W_t = W_0 e^{rt} \quad (1)$$

where W_0 is an initial weight at the beginning of the time interval defined by t; r is the instantaneous relative growth rate, and e is the base of the natural logarithm. With this type of growth curve, a straight line is obtained on a semi-log (log-normal) graph of lnW and t since lnW and t are linearly related as expressed by:

$$\ln W_t = \ln W_o + rt \qquad (2)$$

Figure 4 shows a theoretical growth curve from equation (2). When food is non-limiting, the growth rate is optimal and the growth curve has the slope described by r in equations (1) and (2). If food remains non-limiting, optimal growth continues as illustrated by the dotted line in Figure 4. When food becomes limiting, growth rate slows and weight gain is also slowed. The biomass in the pond at the time of the instantaneous change in the

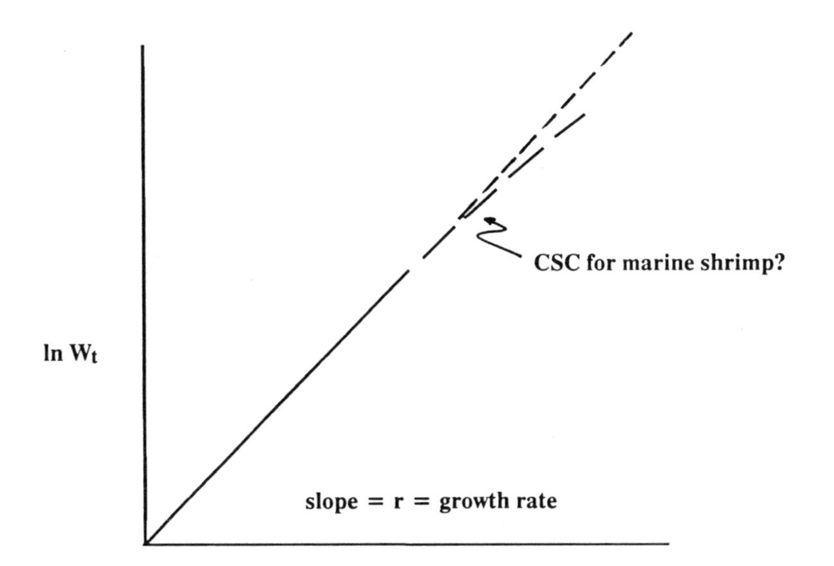

FIGURE 4. Log-normal theoretical curves of the equation $W_t = W_o e^{rt}$. Heavy solid and dotted straight lines define optimal growth, and the curved dotted line represents a change of growth rate indicating that the population biomass has exceeded the critical standing crop, CSC (Hepher 1978), defined as the instantaneous change in the slope, r. This theoretical condition was used as a simple practical means to track the population growth and CSC levels using semi-log plots as illustrated in Figure 5.

growth rate, r, which is also the slope of the growth curve graph of equation (2), defines the critical standing crop, CSC. We used this basic principle to monitor growth in a pond so that supplemental feeding could be instituted or increased once negative changes in growth rate (indicative of the population reaching and exceeding its CSC) were reached.

To monitor growth, a program of pond population sampling was instituted which measured mean size at least once every two weeks. Data points were plotted on semi-log graph paper with ln W (mean size) on the y-axis vs. time in days since stocking on the x-axis. Figure 5 shows a representation of the type of graphs that were plotted using a hypothetical growth curve for a 19 ha (190,000 m^2) pond. For the sake of the example, it was assumed that this pond was stocked at a density of 7 PLs/m^2. Using the Delphi approach it was determined that the pond could be harvested at between 4.0 and 5 months or between 120 and 150 days after stocking. Mean harvest size was determined to be 21 g in the example, but in practice any size can be used as the initial estimator. The first step was to simply place a "terminal" data point at the coordinates on the graph representing the ordinate axis of final mean harvest size (21 g in the case of Fig. 5.) and the two abscissa points of 120 and 150 days. Two lines, drawn from these coordinates to the origin will define an upper and lower expectation of growth. As can be seen from Figure 5, these lines can be used to readily estimate when the expected CSC value of the pond population will be reached if no supplemental feeding is provided.

Using the 300-500 kg/ha/yr extensive production, the CSC of a population was estimated to range between 30 and 50g/m^2. Mortality and overall survival do not play direct roles in this method, since we assumed that survival is not sensitive to density at the densities utilized (3-10 PL/m^2) when feed is not limiting (as presumably it would not be if supplemental feeding was initiated at appropriate times as judged from the shifts in slope of the curve derived from plots of the pond sample data).

Conveniently, the instantaneous population number, N_t, can also be described graphically in the same general way as W_t. In this regard, the population size, N_t, is described as follows:

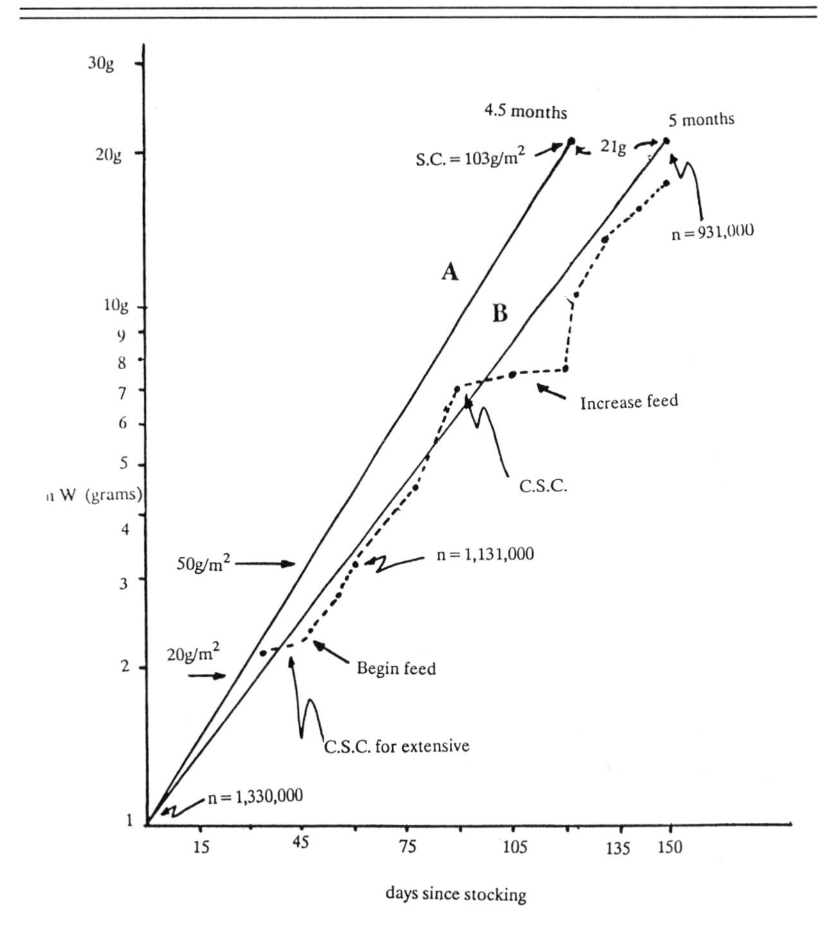

FIGURE 5. Hypothetical curves illustrating the semi-log plots of expected (solid lines) and actual (dotted line) population mean size vs. time, used to track population growth, critical standing crop levels and timing of supplemental feeding institution and increase. Ordinate scale represents ln W_t (in grams) and ln N_t in numbers. Abscissa represents time in days since stocking. Curve A represents population growth over a 4.0-month (120-day) period. Curve B represents a similar pattern over a 5-month (150-day) period. The dotted line represents the type of curve that would result in the plot of actual sample data points of population mean size. Changes in slope are assumed to occur between 30 and 45 days and between 90 and 120 days, indicating that the critical standing crop has been reached and exceeded due to a limitation in food. In the first case feeding was begun, in the second feeding was increased. Assumptions of the figure include a final mean size of 21 g and a survival of 70% with a stocking rate of 7/m² in a 19.2 ha (190,000 m²) pond.

$$N_t = N_o e^{rt} \tag{3}$$

$$\ln N_t = \ln N_o + rt \tag{4}$$

Equations (3) and (4) have forms which are similar to equations (1) and (2), respectively. Therefore the same log-normal graph paper can be used to estimate the instantaneous population number, N_t. N_t can be plotted on the ordinate axis using a different scale. The absolute value of N_t at any time, t, can then be read directly off the graph once an overall assumption with regards to final survival and final population size is made. In the example illustrated in Figure 5, overall survival was estimated to be 70% using the Delphi approach. Therefore at a stocking rate of 7 PLs/m² the total population stocked is 1,330,000 PLs. At final harvest the population number, N_t is 931,000 shrimp. Using the theoretical graph representing the slower growth, at 60 days the population mean size should be approximately 3.5g, N_t = 1,131,000 and the standing crop 50g/m², the upper limit of the CSC in unfed ponds. We assumed that the pond sample data points should define an empirical curve whose negative change in slope reflects the pond biomass (standing crop) exceeding the CSC. Feeding would either commence if the inflected curve was the first change in the curve's slope or would be increased if an inflection in the curve represented a subsequent exceeding of the CSC. A illustration of this is shown by the dotted line curve in Figure 5. This curve represents actual mean population size and is drawn to illustrate the points discussed here. Inasmuch as the second sample data point taken at approximately the 47th post-stocking day lies outside the curve representing the slower growth rate, it was assumed that the CSC had been reached in the pond and feeding was initiated and continued until approximately day 90. The next two subsequent samples, on days 105 and 120, show that a second CSC level had been exceeded and feeding was increased. The overall growth rate responded accordingly.

Initially, our experience during the first year of pond sampling showed that the actual growth curve, in absolute value, fell below the lower limits defined by the theoretical curve, indicating slower growth than predicted. This meant that we quantitatively underestimated the CSC levels. However, qualitative changes in slope could clearly be discerned provided that enough sampling data points were plotted over the time interval. We estimated that a minimum of bi-weekly samples were necessary and that weekly samples are desirable.

3. Synchronization of hatchery production with pond availability.

The first step in this area was to assess the current pond sizes and match them up with hatchery consignments. Table 3 shows the pond areas and stocking requirements for a density of 8 PL/m^2 at the start of the project in September 1986. Each pond (with the exception of pond 4 which was 37 ha) had a PL requirement that was close to our 2.2 x 10^6 PLs/month hatchery production expectation. Pond 4 was considered to be too large to stock effectively so it was subdivided into two units. Ponds 3 and 5 also had PL stocking requirements larger than the hatchery production could furnish per cycle. However, it was decided that these ponds did not warrant subdivision and that they could be stocked at a lower density than 8 PLs/m^2 inasmuch as our strategy was to stock the ponds in the range of 6 - 10 PL/m^2. It was concluded that although the ponds were large by industry standards for semi-intensive shrimp culture, all of the ponds could be effectively stocked in terms of our hatchery production plan.

Table 3. Pond area and PL stocking requirement at a stocking rate of 8 PLs/m^2. Ponds 8 and 9 were part of the farm but were not planned for production.

Status	Pond no.	Hectares	m^2	At 8 PL/m^2 PL requirement x10^6
OK	1	18	180,000	1.40
OK	2	19	190,000	1.52
OK	3	31	310,000	2.48
Subdivide	4	37	370,000	2.96
Subdivide	5	28	280,000	2.24
OK	6	11	110,000	.80
OK	7	13	130,000	1.00
Not usable	8	13	130,000	1.00
Not usable	9	6	60,000	.48

The next step was to conceptually group various ponds into sets that could be stocked with one hatcery cycle's PL consignment. The groupings were initially done without regard to existing shrimp populations within them. The purpose of this was to consider certain combinations of ponds that could be stocked with the PLs from one hatchery cycle. As seen in Table 4, various combinations of ponds can be made and each of these combinations give similar total areas.

Table 4. Groups of ponds that could be used to achieve a given stocking density for a hatchery production cycle of 3 x 10^6 PLs.

Group	Stocking Density PLs/m²	Pond Nos.	Area (ha)	Mean Area (ha)
A	6	3, 4 & 5	50.00	50.00
		1, 2 & 7	50.00	
		1, 2 & 6	48.00	
		4 A,B & 5 or 6	47.00	
B	7	3 & 6	42.00	42.86
		1, 6 & 7	41.00	
		5 & 7	40.00	
		2, 4 A,B or 5	39.00	
C	8	2, 4 A,B or 5	37.00	37.50
		1 & 2	37.00	
		4 A,B & 5	36.00	
		1 & 4 A,B or 5	35.00	
D	9	2 & 7	33.00	33.33
		3	32.00	
		7, 4 A,B or 5	30.00	
		2 & 6	30.00	
		1 & 7	30.00	

In our planning, we assumed that the hatchery production would not exceed 3 x 10^6 PLs/month (modest production for the hatchery's capacity) and grouped the ponds to receive this many stocklings in order that four different densities (6, 7, 8, & 9 PLs/m^2) could be used. For example, suppose one hatchery cycle produced 3 x 10^6 Pls. We could stock these PLs in one of the combination of ponds given in group B in Table 4. That is, ponds 2 and 4A or 4B could be stocked at 7 PLs/m^2 or ponds 1 & 7 could be stocked at 9 PLs/m^2 or ponds 3 and 4A or 4B could be stocked at 6 PLs/m^2 and so on. If hatchery production was below 3 x 10^6 PLs, the table could still be used to determine pond stocking groups, except the density would be lower. If total hatchery production was below 3 x 10^6 PLs and one of the combinations in group A had to be used for other reasons, then the stocking density would fall below 6 PLs/m^2. However, our profit sensitivity analysis showed favorable economic returns in the second year (1988) for stocking densities above 5 PLs/m^2 (see later).

Once the ponds had been conceptually grouped, the next step in the planning process was to construct a time line for the stocking, growout, and harvest schedule for each pond and each hatchery cycle. We initially desired to use off-the-shelf computer management software to do this; however, we did not find the existing programs (Levine 1986) suitable for our purposes because they lacked the capability to chart the specific details necessary for our project. A simple Gantt chart was therefore created by hand. Figure 6 shows this chart for all ponds and hatchery cycles. The project started in September 1986 and those ponds containing shrimp at that time had been stocked with native post larvae (N, in Fig. 6). Some ponds were not in production (pond 1 and 4B, which was to be constructed out of the sub-division of the large pond 4, Table 3). Figure 6 shows that all PLs from each hatchery cycle for the period September 1986 - September 1988 could be assigned parsimoniously to a particular pond in keeping within the 6-10 PL/m^2 stocking range we set for the semi-intensification of the farm.

Figure 6 was also used to tie into the financial analysis described below. In this regard, each harvest (H, in Fig. 6) represents production and income. Hatchery PLs from cycles that could not be stocked into a harvested pond within the same month as the pond was to be harvested were consigned to nursery ponds. In this way the nursery ponds modulated the flow of material (PLs) from the hatchery to the ponds. This is as important as the use of nursery ponds as a means to improve neo-natal survival and reduce opportunity costs of the use of the larger production units for nursery when non-revenue producing but much smaller nursery ponds are more suitable.

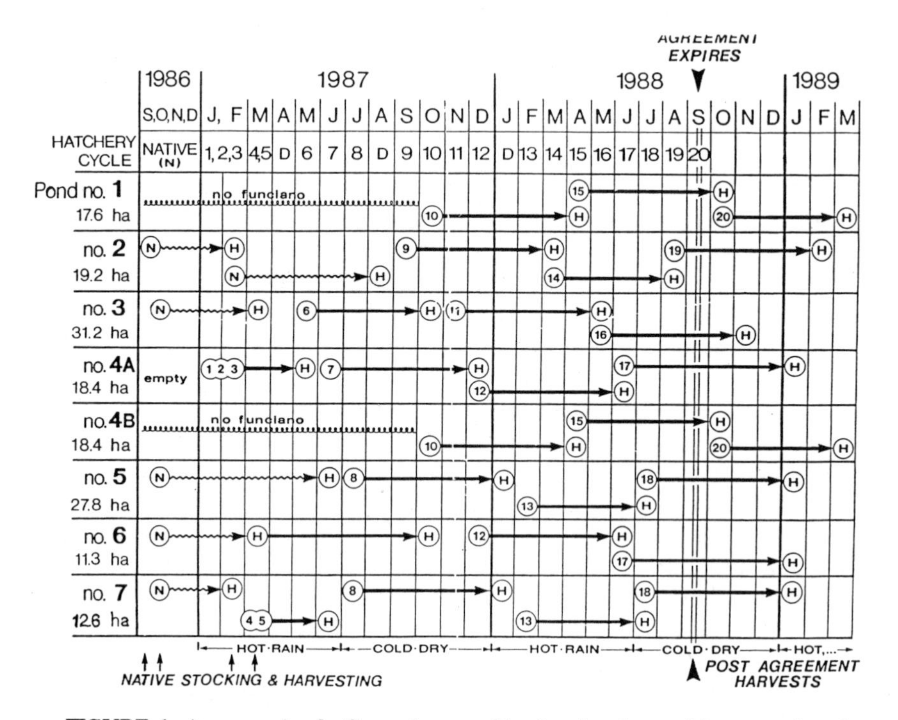

FIGURE 6. An example of a Gantt chart used in planning the stocking, growth and harvesting of each pond in relation to the availability of PLs from each of twenty hatchery cycles for the period September 1986 - September 1988. The period October 1988 through March 1989 is shown to illustrate the "fate" of all twenty hatchery cycles stocked in the ponds. N = native PLs that had been stocked prior to the beginning of the project; H = harvest; circled numbers = stocking of the PL from that numbered hatchery cycle in that particular month, D = "down" time planned into the schedule for hatchery cleaning, repair and to synchronize hatchery production and pond availability. Two growing seasons are identified on the bottom of the chart. Heavy arrows between stocking and harvesting indicators refer to the grow-out period for that pond stocking and vary according to season as described in the text and in Figure 9.

Figure 6 shows the two growing seasons that are present in Ecuador: a hot, rainy season from January through June and a cold, dry season from July through December. The growth of shrimp is affected by this seasonality and was accounted for in the Gantt charting and financial modeling. The length of the solid arrows in Figure 6 is indicative of the length of the growout period for that particular stocking cohort and varies with the time of stocking and the amount of time each growing pond population spends in these seasons.

4. An analysis of the sensitivity of revenues to production variables.

The next, and perhaps most important step in our planning for the semi-intensification of the operation was to estimate whether or not we could make a profit. To do this a financial analysis model was created using Lotus 1-2-3 (Lotus Development Corporation). The approach was simple and direct and consisted of a master spreadsheet model of several sections as shown in Figures 7 and 8. The driver values which drive the main spreadsheet were developed first. The drivers were divided into biological and economic parameters. The former consisted of stocking density, harvest mean size, survival rate and feed conversion ratio (FCR). The economic parameters were the foreign currency exchange rate (sucres to dollars), price of shrimp, feed cost, inflation rate, farm operating costs as a percentage of gross revenue, and tax rates.

The main spreadsheet had three working sections. First, there was a stocking inventory planner whose columns in the spreadsheet corresponded to months of the project, similar to the month designations in the Gantt production time-line chart shown in Figure 6 upon which format the stocking inventory planner section of spreadsheet was based. The row designations contained those headings shown in Figure 8. Up to three stocked ponds could be accounted for in keeping with the groupings shown in Table 3. Depending on the ponds available in that month, the number of PLs needed was determined, which led to a determination of stocking density. Also included in the stocking inventory planner section was estimated growout time, harvest month, estimated yield, selling price for shrimp for that month, and the price of feed. In this regard, we could specify a particular feed cost and selling price of shrimp for particular time periods.

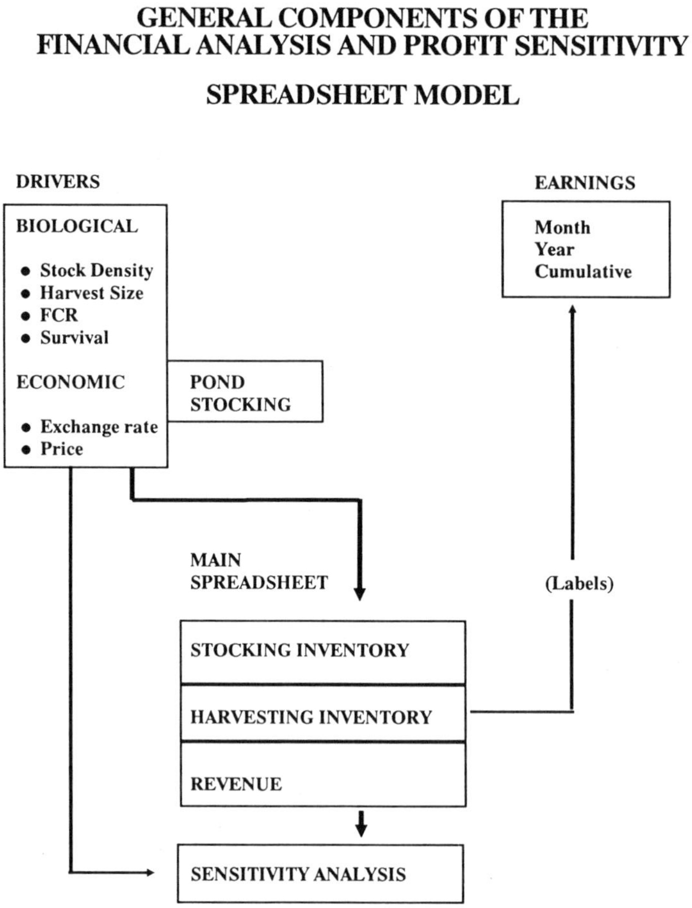

GENERAL COMPONENTS OF THE FINANCIAL ANALYSIS AND PROFIT SENSITIVITY SPREADSHEET MODEL

DRIVERS

BIOLOGICAL

- Stock Density
- Harvest Size
- FCR
- Survival

ECONOMIC

- Exchange rate
- Price

POND STOCKING

EARNINGS

Month
Year
Cumulative

MAIN SPREADSHEET

(Labels)

STOCKING INVENTORY

HARVESTING INVENTORY

REVENUE

SENSITIVITY ANALYSIS

FIGURE 7. Components of the general model used for the financial analysis and revenue sensitivity analysis. Driver parameters "drove" formulae in the main spreadsheet as indicated by the thick arrow. Main spreadsheet consisted of three sections relating to stocking, harvesting and revenue. Sensitivity of profits to various driver parameter values was determined. Output from this was in tabular form and then was converted to graphic form (Figs. 10-13). One stocking density in the driver section could be used to rapidly assess the effect of this parameter on profits. A separate pond stocking driver gave the capacity to input the actual stocking densities as they occurred. Some calculations of the spreadsheet appeared as labels in the earnings section to conveniently visualize the effect on earnings of changing various driver parameter values.

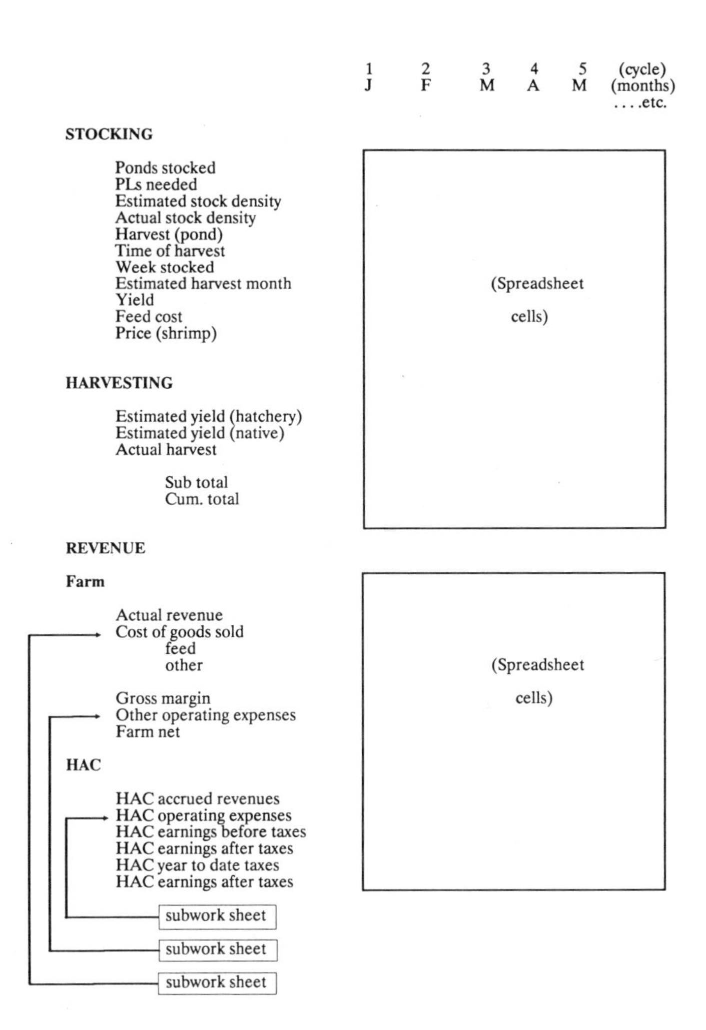

FIGURE 8. General row and column headings of the spreadsheet cells. Estimated and actual densities and yields could be entered in the spreadsheet. Sub-worksheets could be used to input actual costs of the farm operating expenses as cost of goods sold and other operating expenses. Similarly a breakdown of the estimated and actual operating expenses could be entered into a sub-worksheet. This gave precision to the profit sensitivity analysis and output (Figs. 10-13).

The second section consisted of a harvest inventory planner which was tied by formulae to the stocking planner and reflected the production outputs of the system. One of the most important links in the relationship of the stocking inventory planner and the harvesting inventory and revenue sections involved an estimate (programmed into the spreadsheet) of the production growth function. This function described the time period of growth of the cohort stocked in a particular season. We determined this period using a combination of Delphi survey information and a simple model relating shrimp growth with time of stocking at different times within a growth season (either the wet, warm season or the dry, cool season). Such a total growth period for the population was determined depending upon when the two seasons overlapped with three major periods of growth; defined by us as indicative of the three major tangents to the asymptotic growth vs. time curve. Figure 9 illustrates the method.

First we determined, using the Delphi approach, what the growing time would be in the "warm" season and in the "cold" season. The former was determined to the 4.5 months and the latter 6 months. In other words, common knowledge among the current pond managers and other knowledgeable people who had mentally kept track of pond production indicated it took as long as 6 months and as little as 4.5 months to grow a crop under the specific conditions in the area of our ponds. This was the only manner in which "quantitative" data were available to us since there were no such published data for Ecuador.

Second, we assumed that shrimp growth is logarithmic as shown by the curve in Figure 9. We assumed that the final shrimp size is the defined goal and does not change in the different growth scenarios; only the growout time changes. Therefore, harvest size would be reached between 4.5 and 6 months depending upon when in the hot or dry season the pond was stocked.

We estimated 6 growout scenarios. As an example, consider the 4.5 and 6 month periods shown in Figure 9. If a pond was stocked in the beginning of the hot season such that all phases of the growth period could be completed within this season, then the population was assigned a growth period of 4.5 months (scenario 6 in Fig. 9). Conversely, if the entire growth cycle occurred within the cold season, we assigned this population a 6-

SIMPLIFIED GROWTH/SEASON FUNCTION

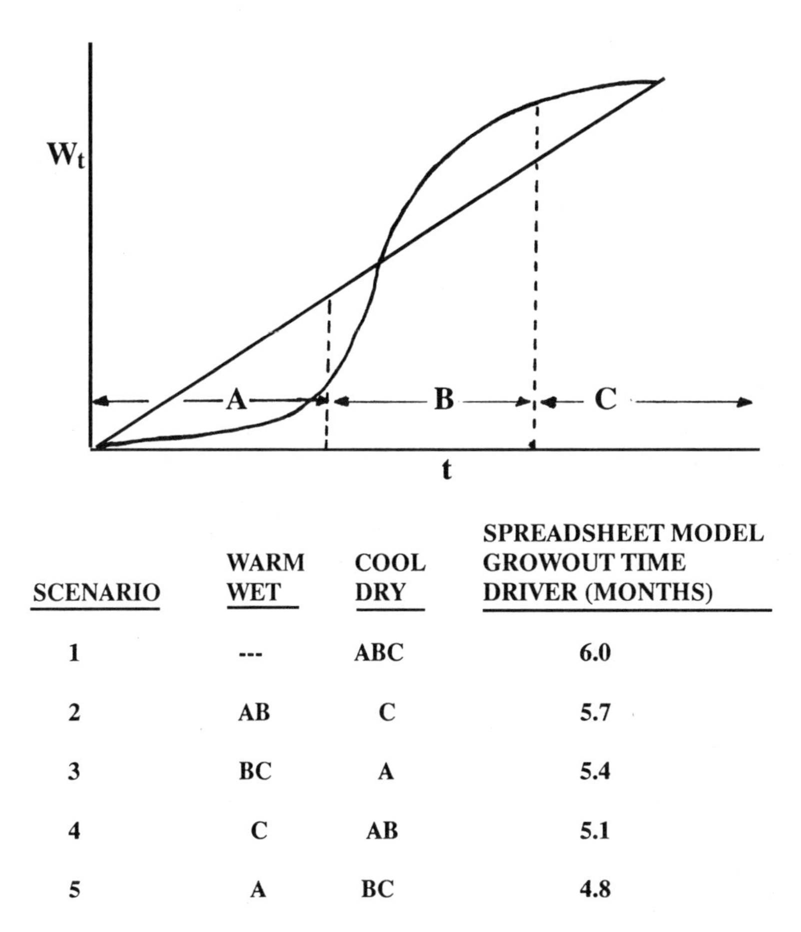

SCENARIO	WARM WET	COOL DRY	SPREADSHEET MODEL GROWOUT TIME DRIVER (MONTHS)
1	---	ABC	6.0
2	AB	C	5.7
3	BC	A	5.4
4	C	AB	5.1
5	A	BC	4.8
6	ABC	---	4.5

FIGURE 9. Simplified pond population growth vs. time curve (top section) separated into three growth periods, A, B & C, approximately corresponding to the major "tangents" to the asymptotic growth curve. Lower section gives six scenarios which result in six different overall growth periods (right hand column) depending upon which combination of growth periods occur in one of two (warm or cool) growth seasons as explained in the text.

month growing season driver value in our spreadsheet (scenario 1 in Fig. 9). To make the model more precise, we assigned four other values to the population growing season depending upon when the growth periods occurred. Each scenario differed by 3 month increments. For example in scenario 2 we assumed if the population had its first two periods, A & B, in the warm season and the final stage C in the cold season, that this would be the next to longest growing season since it is during the last growth period (period C, in Fig. 9) that the most absolute weight gain is realized since growth (i.e., weight gain over a time interval) is a power function of the weight gain at the beginning of the interval (Hepher 1978). Conversely, scenario 5 assigns a 4.8-month growing season to a population whose latter two growth stages are in warmer weather.

The spreadsheet model driver values for growout time took on values determined by which of the six scenarios could be ascribed to the population in question, which, in turn was determined by the calendar month in which the pond was stocked. As actual costs were incurred, actual hatchery PLs produced, and actual pond growth and production experienced, values relating to these were entered into the spreadsheet model. Consequently as the project progressed, the financial analysis consisted more and more of actual costs and revenues in place of projected values. This was reflected in the revenue sensitivity analysis.

The third part of the main spreadsheet consisted of a revenue section. This section estimated gross and net income based on the stocking and harvesting inventory planner and to some degree the values in the driver section. Sub-work sheets illustrated in Figure 8 could be used to calculate cost of goods sold, operating expenses other than deductible from the gross margin, and the operating expenses of our company. From the revenue section of the spreadsheet, calculated values were summarized and recorded as labels in an earnings box (Fig. 7). In this way, the "bottom line" effects of changing the driver values could be conveniently visualized. Figure 7 also shows that calculated values were transferred as labels into the driver box for convenience. Furthermore, the spreadsheet had a pond stocking density driver (Fig. 7) which contained default values for each pond if actual values were not entered into the main spreadsheet.

The fourth part of the spreadsheet consisted of a sensitivity analysis section. In this section, the values for yield, and for gross and net income were listed in tabular form for two years (1987 and 1988). The values were calculated in the spreadsheet for various driver values of one parameter at one value for the other three parameters. In this way, the sensitivity to bottom-line profits (e.g., estimated after tax, EAT, earnings) could be seen for the four parameters: stocking density, food conversion ratio (FCR), harvest size, and survival.

Once the sensitivity tables were produced we used the Lotus 1-2-3 graphics package to produce plots of revenue sensitivity to these parameters. In this way we could conveniently visualize the slope of the plot relating revenues (plotted on the abscissa) to driver values. The degree of positive or negative slope was of interest to us since we wished to target our management to those parameters which affected our revenues the most, as judged by the steepness of the slope of the graphs.

Figures 10, 11, 12 and 13 show the plots of four estimated revenue levels for the four parameters for two years 1987, and 1988. The revenue levels were our company's break-even level (HAC B/E in the figures), farm net revenue obtained after accounting for operating expenses paid for by our Ecuadorian principals, subcontractor gross revenue obtained as a percentage disbursed to our subcontractors who provided the management and technical assistance, and finally our estimated after-tax revenue (EAT).

A comparison of the results between years for each of the analyses shows that, as expected, the net profit to the farm for 1987 was acceptable, but our revenue as a percentage of this was below break even levels. However, this situation changed dramatically for 1988 as seen in the analyses for this year. The operation showed favorable economic returns for all levels of the parameters examined. It must be noted that the 1988 values take into account the profit or losses incurred in 1987 which are carried forward in the spreadsheet to 1988.

The curve of our earnings after taxes crosses the break-even curve at conservative values well within those of the management plan we hoped to carry out. These cross-over points are approximately 5 PL/m^2 for stocking density (Fig. 10), and 14 grams for mean harvest size (Fig. 13). In the case of food conversion ratio, the break-even and after-tax earning curves do not "cross" within the levels defined in our analysis, Figure 12. This indicates

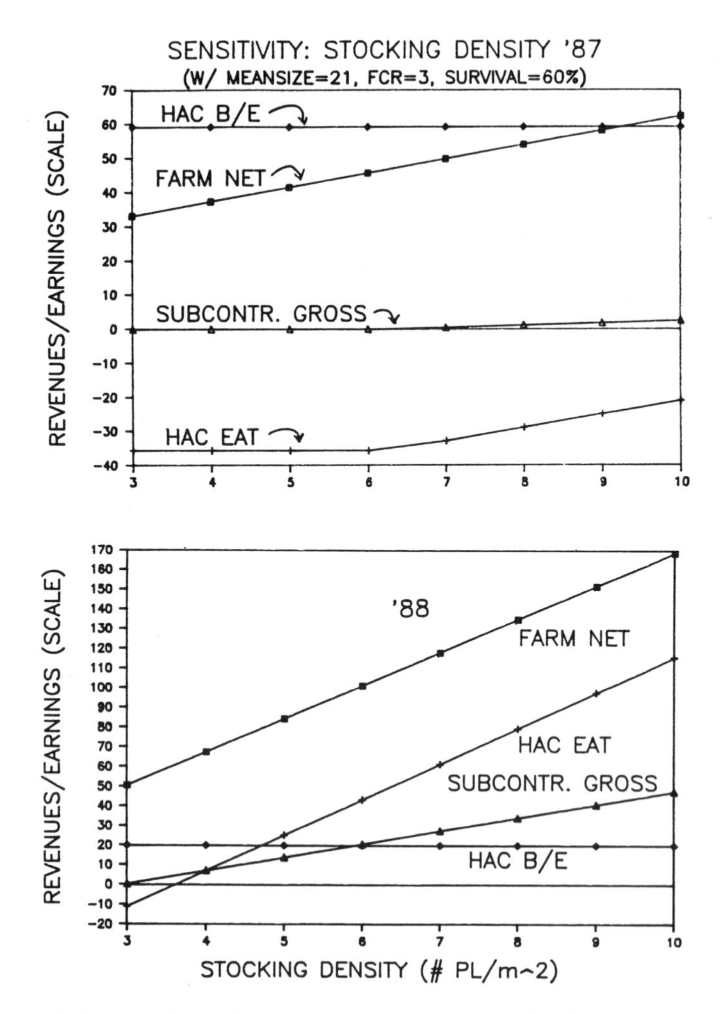

FIGURE 10. Sensitivity of revenues to stocking density for 1987 (top frame) and 1988 (bottom frame). Harvest mean size is assumed to be 21 g, the food conversion ratio (FCR) to be 3:1, and survival to be 60%. HAC B/E = break-even levels for the Hawaii Aquaculture Company equity share of the project; FARM NET = gross margin net revenue for the farm not considering the HAC costs; HAC EAT = estimated after tax earning of the Hawaii Aquaculture Company; SUBCONTR. GROSS = gross revenue to technical managers subcontracted by HAC. Revenues and earnings are expressed in scaled relative (not absolute) values in thousands of dollars. The actual dollar value for the revenues and earnings are not shown.

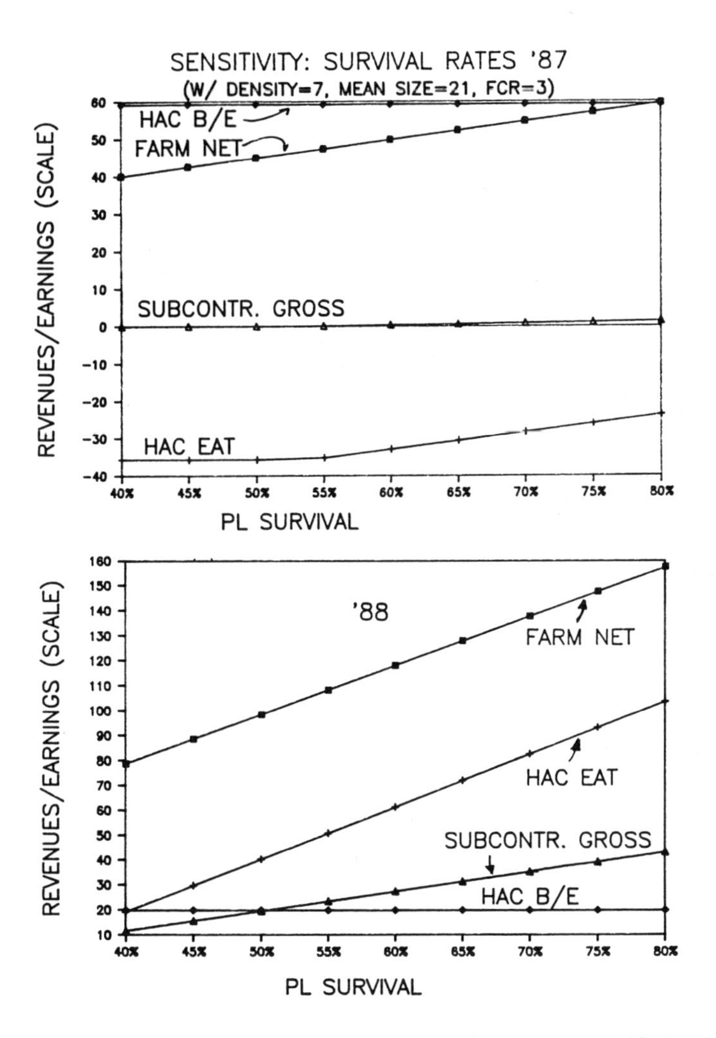

FIGURE 11. Sensitivity of revenues to survival rate for 1987 and 1988. Stocking density is assumed to be 7 PLs/m^2, all other parameters and terms are as described for Figure 10.

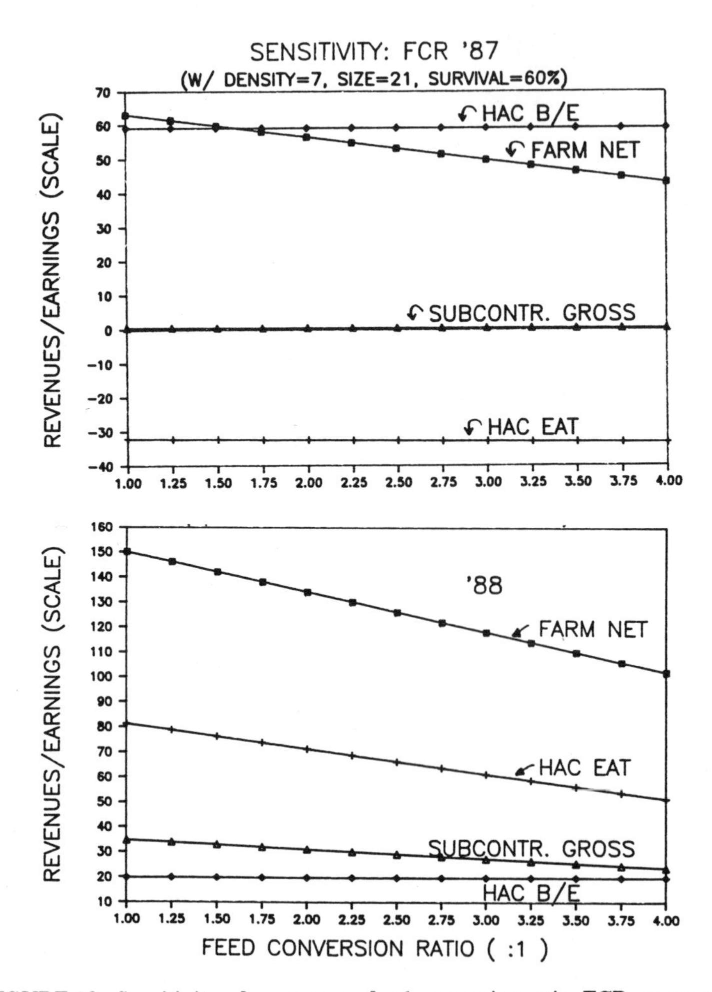

FIGURE 12. Sensitivity of revenues to food conversion ratio, FCR, measured as the ratio of total amount of feed used to total weight of shrimp produced at harvest. Survival is assumed to be 60%; all other parameters and terms are as described for Figure 10.

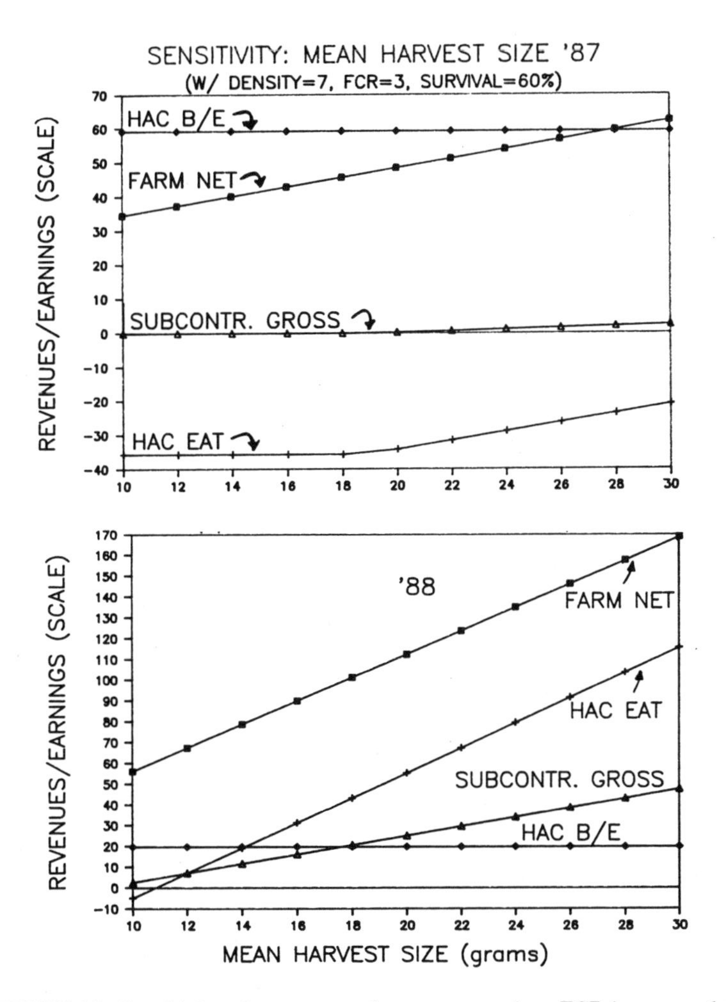

FIGURE 13. Sensitivity of revenues to harvest mean size. FCR is assumed to be 3:1. All other parameters and terms are as described for Figure 10.

that FCR levels greater than 4:1, the highest (i.e., poorest) value plotted in Figure 12, can be obtained before the EAT drops below the break-even levels. In the case of survival (Fig. 11), the cross-over point is at the 40% level, the lowest plotted.

The next comparison we wished to make in designing a strategic plan for the project was to decide which parameter or combination of parameters are most important to our revenue. An examination of the steepness of the EAT curve for all the analyses shows that revenues are approximately equally sensitive to final mean size, survival, and stocking density. Moreover, this sensitivity is large - the slopes indicate sharp rises in profits between unit increases of the parameter. In the case of FCR, profits were not as relatively sensitive to this parameter (Fig. 12). Therefore, we concluded that feed costs as reflected in the FCR were not relatively as important as other costs. We therefore did not attempt to increase or decrease FCR by any managerial intervention or other means.

Regarding the other parameters, we felt that survival was not as much within our control as harvest size and stocking density, which we could manage directly. Final harvest size could be "managed" by increasing the growing time, allowing the shrimp population to reach a larger mean size. However, in many cases this extension would conflict with other farm needs such as for cash flow or nursery space. We decided therefore to concentrate our management of revenues through pond stocking density. We had planned a conservative hatchery production per month (2.2×10^6 PLs) which we felt could be increased using the existing hatchery capability. It was important to keep all ponds stocked at densities greater than 5 PL/m^2, and we felt that 10 PLs/m^2 would be an upper limit we wished to use until we had a better understanding of pond growth and CSC values in relation to feeding schedules. We planned our management accordingly.

5. Reduction of risk from low dissolved oxygen using variable stocking densities and harvest sizes.

The large ponds in this present project (Table 3) presented a potential risk due to low dissolved oxygen (DO). Losses, such as disease-related death in the case of hatchery tanks or mortality related to low dissolved oxygen in the case of ponds, occur in the units themselves regardless of size.

Just as variation in hatchery production is production unit-specific, so is variation in pond production specific to the pond as a production unit. The smaller the number of units there is in a system the greater risk this presents regarding acute losses. In our case we had only seven ponds and we knew the biological oxygen demand (BOD) in the incoming water could be high. This places ponds at risk due to low DO, especially under a supplemental feeding regimen.

Aerating smaller ponds is no problem, but aerating these large ponds was not technically or economically feasible. We therefore planned to "manage" low dissolved risk by manipulating stocking and harvesting. Our strategy was based upon three interrelated areas: the respiratory physiology of shrimp like *P. vannamei*, the relationship of a shrimp's surface-to-volume ratio to its risk of death due to low DO, and the relationship of selling price to shrimp size.

Penaeus vannamei, like most benthic detritivorous crustaceans, is an "oxyconformer" (Cameron and Magnum 1983). It will regulate its intake of oxygen through its ventilation rate (and other means) and to some degree its physiological oxygen requirement to the ambient dissolved oxygen level. However, there is a threshold level of ambient dissolved oxygen below which an individual will die. This threshold level is related to an animal's surface-to-volume ratio. Smaller animals withstand lower ambient dissolved oxygen better than larger ones; since the body surface-to-volume ratio of small animals is higher than larger animals, small animals are afforded a greater opportunity to absorb across their gill surface the oxygen required to satisfy the physiological demand of the body volume.

In the presence of declining ambient dissolved oxygen levels, larger animals are the first to die and therefore are at greater risk. We used this relationship to design a strategy to minimize risk to shrimp pond populations by stocking larger numbers of PLs and harvesting sooner when animals are smaller and at less risk. Obviously there are economic implications to this action since smaller animals command lower market prices. Figure 14 shows, however, that similar economic returns can be realized at two different stocking densities. Low stocking density leads to larger, higher priced shrimp who are at greater risk to low dissolved oxygen. The higher number of smaller shrimp produced are economically equivalent to fewer

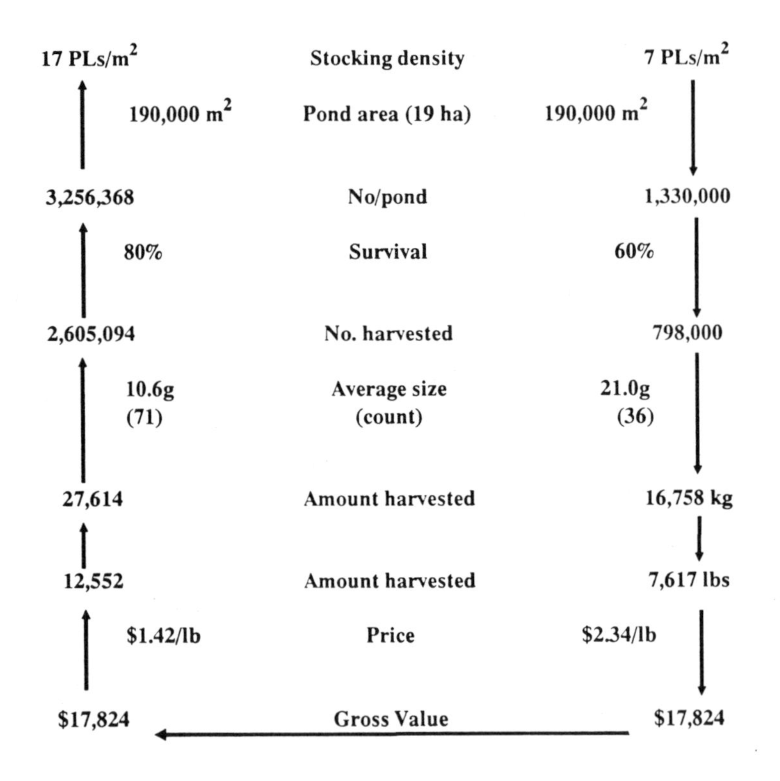

FIGURE 14. Flow chart showing the economic equivalency of two stocking densities when animals are harvested at larger sizes (36/40 count) for a higher price when stocked at a lower density 7 PL/m² and when animals are harvested smaller (71/80 count) for a lower price. A strategy of using higher stocking densities can be employed to harvest smaller animals which are at less risk due to low ambient dissolved oxygen (DO) levels in the pond.

large shrimp. The economic returns are equivalent, but the risk is not. If low dissolved oxygen losses occur, this would reduce the value of the crop even further.

The increase in PL requirements from 7/m^2 to 17/m^2 to stock in the higher scenario would not put an appreciable burden on hatchery production. Lower stocking densities (15 PL/m^2) for one pond may be more realistic, and we assumed that the lowered revenues from stocking at this level and harvesting at the small size (71/80 count) shown in Figure 14 could probably be absorbed. We also assumed that survival would be higher for the smaller shrimp, harvested earlier. Also, the estimated growout time for the smaller ("low" risk) group would be shortened to approximately 100 days. For the larger animals the growout period would be approximately 135 days.

CONCLUSION

A simple, direct, and cost-effective strategy utilizing basic graphing methods, off-the-shelf computer software, and the conceptual framework of methodologies developed for other industries, can be applied to developing commercial marine shrimp and other aquaculture projects. Such a strategy allows project principals, consultants and advisors to rapidly assess the feasibility and profitability of the project.

LITERATURE CITED

Cameron, J.N. and C.P. Magnum. 1983. Environmental adaptations of the respiratory system: ventilation, circulation, and oxygen transport. *in* D.E. Bliss, editor. Biology of Crustacea, Academic Press, N.Y.

Fusfeld, A.R. and R.N. Forster. 1971. The Delphi technique: survey and comment. Business Horizons 14(6):63-74.

Griffin, W.E., W.E. Grant, R.W. Brick and J.S. Hanson. 1984. A bioeconomic model of shrimp mariculture systems. Ecological Monographs 25:47-68.

Hepher, B. 1978. Ecological aspects of warm-water fish pond management. Pages 447-468 *in* S.D. Gerking, editor. Ecology of freshwater fish production. John Wiley & Sons, New York, USA.

Jessup, P.T. 1985. Continuing process control and process capability improvement. A guide to the use of control charts for improving quality and productivity for company, supplier and dealer Activities. Statistical Methods Office, Operations Support Staffs, Ford Motor Company, Sept. 1985.

Johns, M., W. Griffin, A. Lawrence and J. Fox. 1981. Budget analysis of penaeid shrimp hatchery facilities. Journal of the World Mariculture Society 12(2): 305-321.

Lawrence, A., J.P. McVey and J.V. Huner. 1985. Penaeid shrimp culture. *in* J.V. Huner and E.E. Brown, editors. Crustacean and mollusk aquaculture in the United States. AVI Publishing Co., Inc., Westport, CI.

Levine, H.A. 1986. Project management using microcomputers. Osborne McGraw-Hill, Berkeley, CA.

Linstone, H.A. and M. Turoff. 1975. The Delphi method. Addison-Wesley, Reading, Mass. USA.

Mock, C. 1980. Trip report by C.R. Mock to Sea Farms de Honduras April 11-18, 1980. National Marine Fisheries Service - SEFC, Galveston Laboratory, Galveston, Texas.

Mock, C. 1981. Report on penaeid shrimp culture consultation and visit, Guayaquil, Ecuador, South America, and Panama, Central America, August 12 to September 20, 1981. National Marine Fisheries Service - SEFC, Galveston Laboratory, Galveston, Texas.

SEAFDEC, 1984. A guide to prawn hatchery design and operation. South East Asian Fisheries Development Center (SEAFDEC), Iloilo, Philippines. Aquaculture Extension Services Manual No. 9.

Zuboy, J.R. 1981. A new tool for fishery managers: the Delphi technique. North American Journal of Fisheries Management 1:55-59.

PROSPECTS FOR THE APPLICATION OF BIOTECHNOLOGY TO THE DEVELOPMENT AND IMPROVEMENT OF SHRIMP AND PRAWNS

Dennis Hedgecock and Spencer R. Malecha

ABSTRACT

Forecasting applications of biotechnology to the breeding of shrimp and prawns for aquaculture must be based on a realistic assessment of the current status of husbandry and genetic management for these undomesticated crustaceans. There are, for example, no pedigreed stocks or breeds of shrimps or prawns under development. Since shrimp and prawn aquaculture does not even utilize the "low technology" of traditional animal breeding, it is difficult to be very optimistic about the prospects for the application of "high biotechnology" in the near term. Nevertheless, while discouraging a view of biotechnology as a shortcut to the breeding of a "super shrimp," we believe that biotechnology might be employed in concert with developments in basic husbandry and culture in the following areas: (1) management of brood stock by tracking of maternal and paternal lineages using mitochondrial and nuclear DNA typing or fingerprinting; (2) control over sex determination, sex reversal and use of mono-sex culture; (3) rapid and precise diagnosis of molecular defects or pathogens for efficient prophylaxis; (4) genetic and endocrinological control of growth and reproduction; and (5) product identification. Recent advances that simplify or automate molecular biological methods such as enzymatic amplification of DNA ("molecular cloning"), DNA sequencing, and probing with non-radioactively labeled oligonucleotides, have vastly expanded the utility of biotechnology in these areas. This paper reviews and forecasts the application of biotechnological techniques in the five areas mentioned above and discusses the advantages and limitations of particular techniques.

INTRODUCTION

Advances in molecular biology and their application in biotechnology and genetic engineering have been widely proclaimed by both the scientific and popular presses. Developments in these areas are indeed exciting but their impact on agriculture in general, and on animal improvement in particular, may be much more subtle and felt over a much longer period of time than the public has been led to believe. The feasibility of inserting foreign genes into farm animals has been demonstrated (Hammer et al. 1985), but its role in improving production is likely to be ancillary to traditional breeding methods in the foreseeable future (Smith 1988; Crittenden and Salter 1988). However, molecular methods will make traditional breeding much more efficient, mainly by providing a much larger number of genetic markers linked to genes that control economically important traits affecting production (Edwards et al. 1987; Soller and Beckmann 1988; Stuber 1989).

How will biotechnology affect shrimp and prawn aquaculture[1] in the coming decade? An attempt to answer this question forces a critical comparison of the states of brood stock management in crustacean aquaculture and traditional livestock agriculture. Clearly, shrimp and prawn aquaculturists have not mastered even the "low" technology of either domestication or traditional animal breeding methods. It is likely, therefore, that application of biotechnology to crustacean aquaculture will be slower than in traditional livestock agriculture. An analogy with human motor development — "we must learn to walk before we learn to run" — seems apt. With respect to biotechnology, shrimp and prawn aquaculture are at the "crawling" stage. Nevertheless, just as in agriculture, new molecular methods can and should be employed in concert with much needed developments in basic husbandry and brood stock management. From this perspective, we provide an overview of some molecular methods and discuss their likely applications to the development and improvement of shrimp and prawn brood stocks in the coming decade.

---

1 We use the words "shrimp" or "marine shrimp" to refer to members of the family Penaeidae and the words "prawn" or "freshwater prawn" to refer to members of the genus *Macrobrachium*, chiefly *M. rosenbergii*.

TECHNOLOGY FOR TRADITIONAL BREEDING NEEDS TO BE DEVELOPED

Lack of development of traditional breeding methods for marine shrimp or freshwater prawns has been discussed in detail elsewhere (Malecha 1983; Malecha and Hedgecock 1989). The vast majority of the world's shrimp and prawn aquaculture relies on collection of wild brood stock, and no genetic improvement or domestication is possible in these "open" brood stock populations. Even where brood stock populations have been isolated or semi-isolated from natural populations (an important prerequisite for domestication), there is reason to be concerned about the genetic health of these "closed" brood stock populations. Indeed, genetic variability may be lost from them owing to random sampling errors in the transmission of genes from generation to generation in small populations, a process called genetic drift. The amount of genetic drift is directly related to the effective size (N_e) of a brood stock, not its numerical census. N_e is usually lower than the numerical census of brood stock for a variety of reasons (Fig. 1A). Inequality between the numbers of males (M) and females (F) that actually mate and produce offspring reduces N_e relative to the total number of brood stock, N = M+F, according to the formula:

$$N_e = 4MF/(M+F)$$

As illustrated in Figure 1A, a "breeding population" of 1 male and 20 females will have an N_e of 3.8, not 21.

Another factor that increases the chance loss of genetic variation and appears to be particularly important in highly fecund aquatic species is inequality in the reproductive contributions of individual brood stock to subsequent breeding populations (Malecha and Hedgecock 1989). The effective size of a brood stock population, N_e, as a function of breeding population size, N, and variance in the number of offspring per parent, V_k, is given by:

$$N_e = (4N-4)/(V_k+2)$$

Figure 1B illustrates this point with diagrams representing egg masses

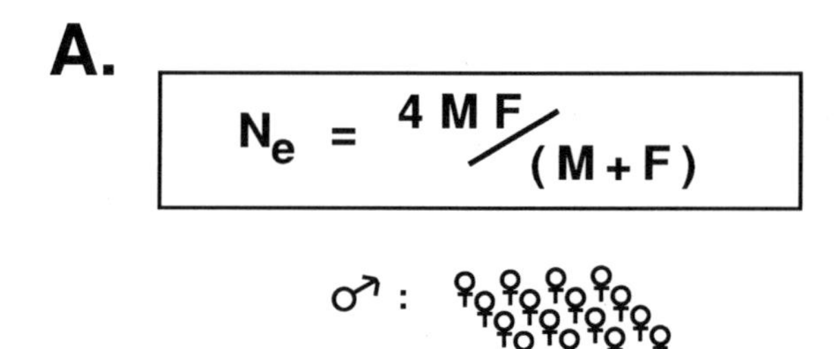

1:20

$N_e \approx 3.8!$

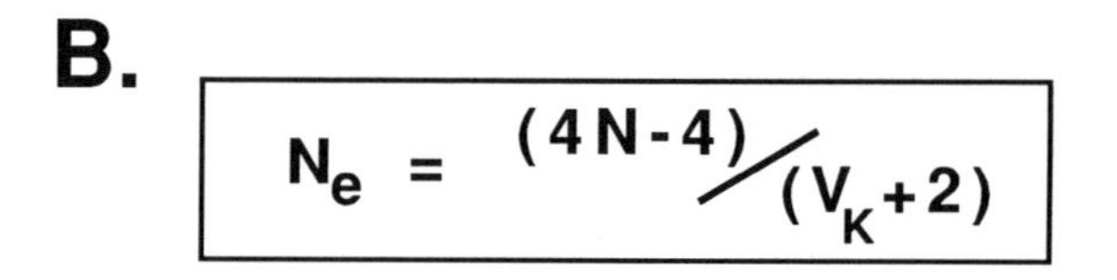

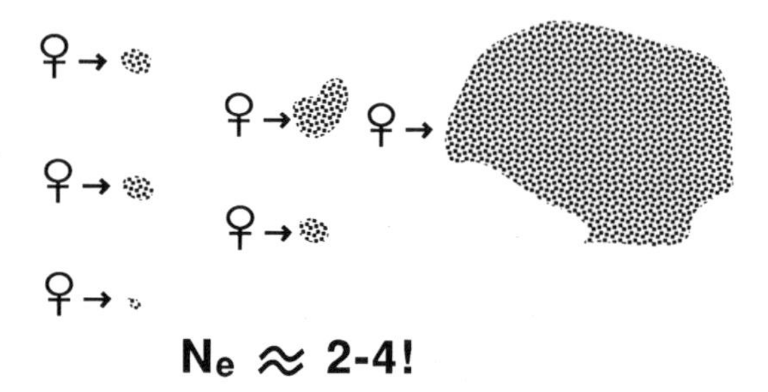

Figure 1. Effective population size. A. Effect of unequal numbers of breeding females (F) and males (M) on effective population size, N_e. Breeding population is shown as 21 individuals, 1 male and 20 females, yet the effective size is 3.8. B. Effect of the unequal contribution of progeny numbers — described as V_k (the variance of offspring number reaching reproductive age) — on N_e. The stippled areas represent brood size (either egg number or larvae). Since the exact brood size is not known, an "estimate" of N_e is given as approximately 2-4.

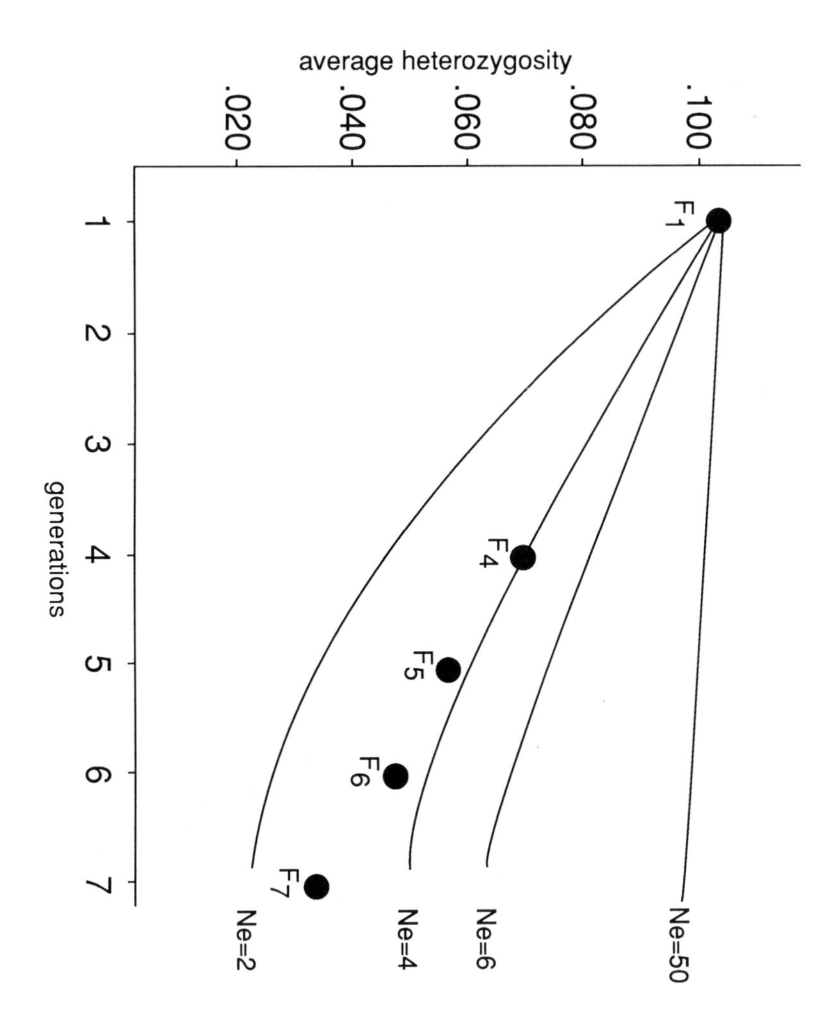

Figure 2. Brood stock heterozygosity. Observed values of the average heterozygosity of allozyme loci in seven generations of ***Penaeus japonicus*** brood stocks (F1 - F7) compared with theoretical values of heterozygosity levels calculated in brood stock populations of N_e = 50, 6, 4, and 2. After seven generations, an aquaculture brood stock censused at 600 individuals had an apparent effective size between 2 and 4 (redrawn from Sbordoni et al. 1987).

or larval broods of unequal size produced by six female parents each presumably mated to a different male (N=12). The unequal contribution of each female to the next generation reduces N_e to a much smaller number than 12, probably approximately 2-4, depending upon the actual variance, V_k.

The scant evidence that exists regarding isolated hatchery populations of marine shrimp indicates that the inequality of reproductive contributions is large (Kawagashi et al. 1986) and that it may be the cause of losses of genetic variation (Sbordoni et al. 1987). The latter authors have shown that a brood stock population of 600 *Penaeus japonicus* that was propagated by an Italian hatchery through seven generations lost genetic variability (measured as average heterozygosity of allozyme loci) as if its effective size was between 2 and 4 (Fig. 2). A discrepancy between actual and effective population sizes has recently been documented for hatchery stocks of Pacific oysters (Hedgecock and Sly 1990) and appears to be quite pervasive in aquaculture.

The inbreeding that inevitably accrues in brood stocks of small N_e can reduce fitness and productivity. For example, Sbordoni et al. (1987) also observed a concomitant decline in the percentage of eggs successfully hatching from the stock of *Penaeus japonicus* that they studied. Biotechnology will not save such inbred stocks, but adoption of traditional animal breeding practices could prevent inbreeding from occurring in the first place. The most important improvement that shrimp and prawn aquaculturists can make in current husbandry practices is to pedigree their brood stock (i.e. trace lines of descent), one of the most fundamental of animal breeding practices. Pedigrees would allow the shrimp breeder to even out the reproductive contributions of brood stock, to prevent the mating of close relatives by chance, and to assess the breeding value of individuals. Development of pedigreed stocks is necessary for practicing traditional selective breeding, which must precede the application of biotechnology or genetic engineering.

To our knowledge, no one is presently developing pedigreed brood stocks of shrimp or prawns. The facilities needed for genetics research and stock development do not exist, and no government support of long-term research is available (Malecha and Hedgecock 1989). Because of this, prospects for applying traditional animal breeding technology to shrimp and prawn aquaculture appear dim.

HIGH-TECHNOLOGY APPROACHES TO TRADITIONAL BREEDING PROBLEMS

Problems that arise in the development of traditional animal breeding technologies, such as identification of individuals and lines of descent in brood stock populations, or control over reproduction and sex, can now be addressed by powerful methods of modern molecular, cell and organismal biology. In this section, we wish to focus attention on some of these methods and their immediate applicability to traditional breeding problems. We will discuss prospects for direct genomic intervention (the popular concept of genetic engineering) in the next section.

Genetic Markers

Advances in biochemical and molecular methods have greatly increased our ability to visualize variability among individuals for specific gene products and for genes themselves. These genetic differences among individuals serve as "markers," the uses of which in aquaculture have been discussed previously (Hedgecock et al. 1976; Hedgecock 1977; Moav et al. 1978). Some of the more important uses considered here are; (1) species identification; (2) pedigree confirmation and analysis; (3) measurement of inbreeding; (4) marker-assisted selection; and (5) product identification.

Allozymes

Protein electrophoresis of genetically encoded enzymes (or allozymes), was the first molecular method to have broad applications to aquaculture (Hedgecock et al. 1976; Hedgecock 1977; Moav et al. 1978). Figure 3 illustrates a commonly observed pattern of allozyme variation. Compared with other animals, decapod crustaceans have rather low levels of allozyme variation (Nelson and Hedgecock 1980; Hedgecock et al. 1982); both freshwater prawns (Hedgecock et al. 1979) and penaeid shrimp (Lester 1979; Mulley and Latter 1981 a,b; Lester 1983; Sbordoni et al. 1987) fit this general pattern. Even low levels of variation can prove useful, however.

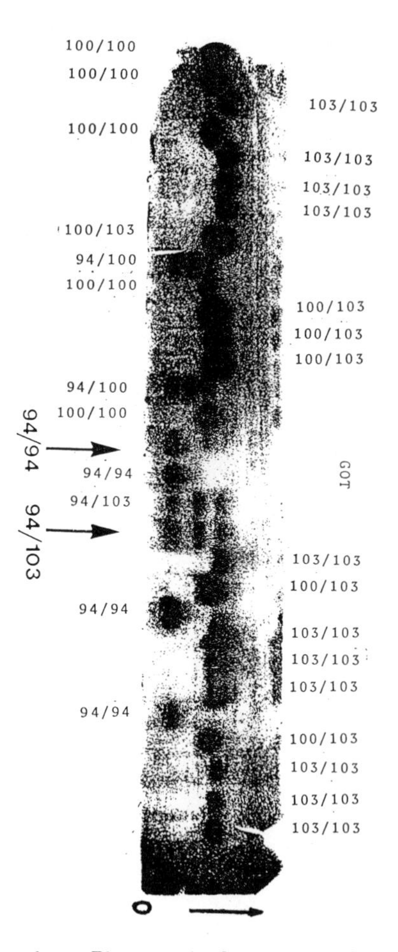

Figure 3. Allozyme markers. Photograph of a starch gel following electrophoresis and staining for the allozyme GOT (glutamate oxaloacetate transaminase) of tissue extracts from 30 freshwater prawns (Zacarias 1986; Zacarias et al. 1990). o="origin," location of the placement of the filter paper strips soaked with the sample homogenate. The arrow indicates the direction of protein migration during electrophoresis. The designations "94," "100," and "103" refer to the alleles controlling the allozyme genotypes. There are three homozygous genotypes, 94/94 100/100 103/103, and three heterozygote genotypes shown among the 30 samples. Two samples, representing the heterozygote 94/103 and the homozygote 94/94, are highlighted. The stained zone in the middle of the heterozygous patterns is a hybrid zone representing a heterodimeric protein.

Previously undescribed subspecies of prawns and shrimp have been identified (Hedgecock et al. 1979; Lester 1983), and allozyme markers in freshwater prawns have been used to demonstrate both the reproductive success of runt males (Telecky 1982, 1984) and the post-larval origins of heterogeneous growth rates (Zacarias 1986; Zacarias et al. 1990). Nevertheless, the few allozyme markers available in shrimp and prawns limits their usefulness in pedigreeing and selection programs. Newer methods for studying variation in deoxyribonucleic acid (DNA) promise to reveal much more extensive variation among individuals.

Mitochondrial DNA (mtDNA)/RFLP

The cells of higher organisms contain mitochondria (tens to thousands, depending on the tissue) that possess multiple copies of a small (usually ≈ 16,000 nucleotide base pairs) circular DNA molecule that encodes ribosomal RNA, transfer RNAs and several of the proteins involved in oxidative phosphorylation, the chief energy-generating function of this organelle. Because of its small size, ease of isolation, maternal inheritance, lack of recombination and intra-individual variation, and rapid nucleotide sequence evolution despite overall conservation of gene order, mtDNA has been the object of evolutionary and population genetic studies for about a decade (Brown 1983; Avise 1986).

The major method for revealing differences among individuals in mtDNA sequence is enzymatic digestion of the molecule followed by gel electrophoresis of the resulting fragments (Ferris and Berg 1987; Gyllensten and Wilson 1987). The overall process is illustrated in Figure 4. First, the mtDNA is isolated from nuclear DNA (Fig. 4A). Then, the mtDNA is incubated with bacterially derived restriction endonucleases which recognize short, specific nucleotide sequences (4 or 6 base pairs for the most commonly used enzymes), and cleave the double-stranded DNA (Fig. 4B). The resulting fragments may then be either radioactively or chemically labeled, separated, and identified on the basis of size by polyacrylamide gel electrophoresis (Fig. 4C). Differences among individuals in the number and sizes of fragments are interpretable as differences in the nucleotide sequences of restriction enzyme recognition sites (Fig. 4D); hence, such differences among individuals are called "restriction fragment length polymorphisms" or RFLPs.

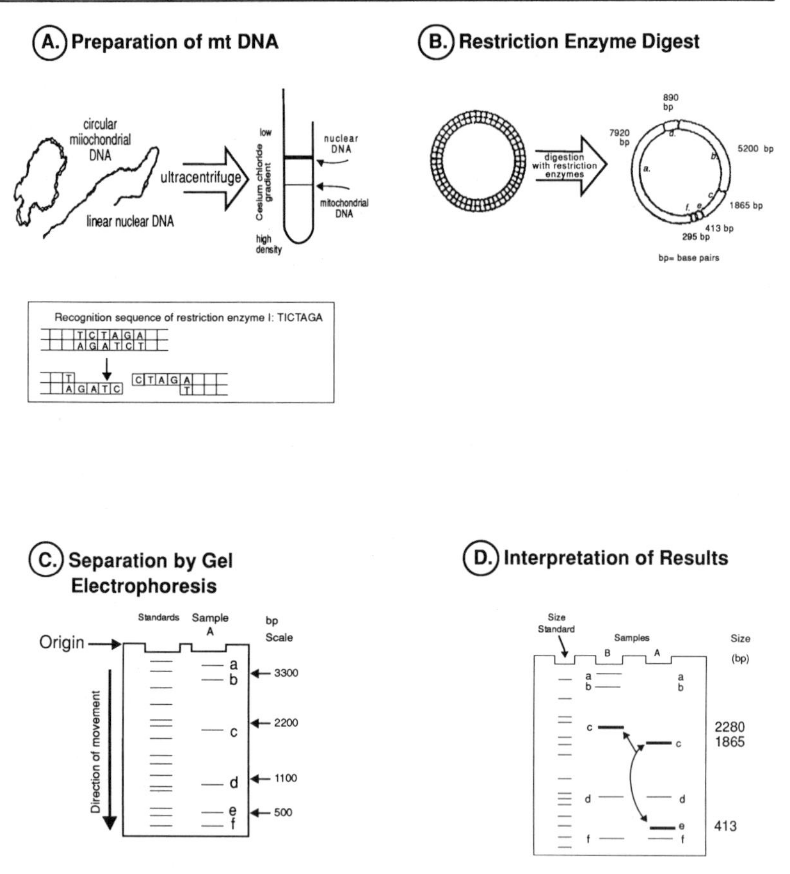

Figure 4. Restriction fragment length polymorphisms (RFLPs). The processes involved in revealing RFLPs in mitochondrial DNA (mtDNA). A: Preparation of the circular mtDNA by ultracentrifugation which separates it from nuclear DNA. B: Cleavage of the circular mtDNA into pieces by a restruction enzyme which recognizes "sites," here illustrated as the DNA sequence TCTAGA and its complementary DNA sister-strand sequence AGATCT as shown in box insert. Following digestion of the mtDNA, six fragments of 7920, 890, 5200, 1865, 413, and 295 base pairs (bp) result. C: The mtDNA fragments are separated by gel electrophoresis (sample A) and compared with a standard. The six fragments (a,b,c,d,e,f) that resulted from the restriction enzyme digestion appear as bands as shown. A DNA base pair (bp) scale is shown on the right. D: When two samples (A and B) are compared, they can be distinguished from one another by band position differences (c and e) which indicate that the mtDNA is genetically different, i.e. "polymorphic" between the two samples (redrawn with permission from Ferris and Berg 1987).

MtDNA RFLPs serve as markers of maternal lineages which allows the analysis of the important problem of variation of reproductive success among individual brood stock. Indeed, Palumbi (1988) has reported loss of mtDNA variation (i.e., loss of maternal lineages) in a closed population of *Penaeus stylirostris*. Such markers might also serve as a means for genetically "branding" a commercially developed brood stock for proprietary purposes.

Nuclear DNA/VNTR

Variation in nuclear DNA can be classified into two types, variation in nucleotide sequence and variation in length. Until very recently, the task of cloning and sequencing specific nuclear genes was formidable. With the advent of the new methods for molecular cloning, however, it has become feasible to gather data on sequence variation in populations.

Variation in both nuclear DNA and mitochondrial DNA length can be revealed by the RFLP method discussed above. The situation with nuclear DNA, however, is quite different than with mtDNA. Scattered throughout the nuclear genomes of most higher organisms are families of related nucleotide sequences that are each composed at any one site of a tandemly repeated, characteristic, small core sequence (Jeffries et al. 1985). Because, in different individuals, each of the sites comprising such a family of sequences may be occupied by different numbers of the core sequence, they are called "variable number tandem repeat" (VNTR) loci. All of the VNTR loci for a particular core-sequence family are revealed by cross hybridization of gel-separated restriction enzyme fragments with a radioactive probe for the core sequence of the family. The resulting array of DNA fragments is essentially unique to the individual and has been called a "DNA fingerprint." This is illustrated in Figure 5.

While DNA fingerprinting may prove valuable in determining the paternity of a brood or the identity of individuals from a commercial stock, the complex inheritance of the DNA fingerprint limits its utility in solving other problems of brood stock management. Specifically, it is difficult to determine which bands in a DNA fingerprint are allelic (alternatives at the same site in the genome), and which are encoded at different sites in the genome; hence, it is difficult to determine whether a band shared by individuals in-

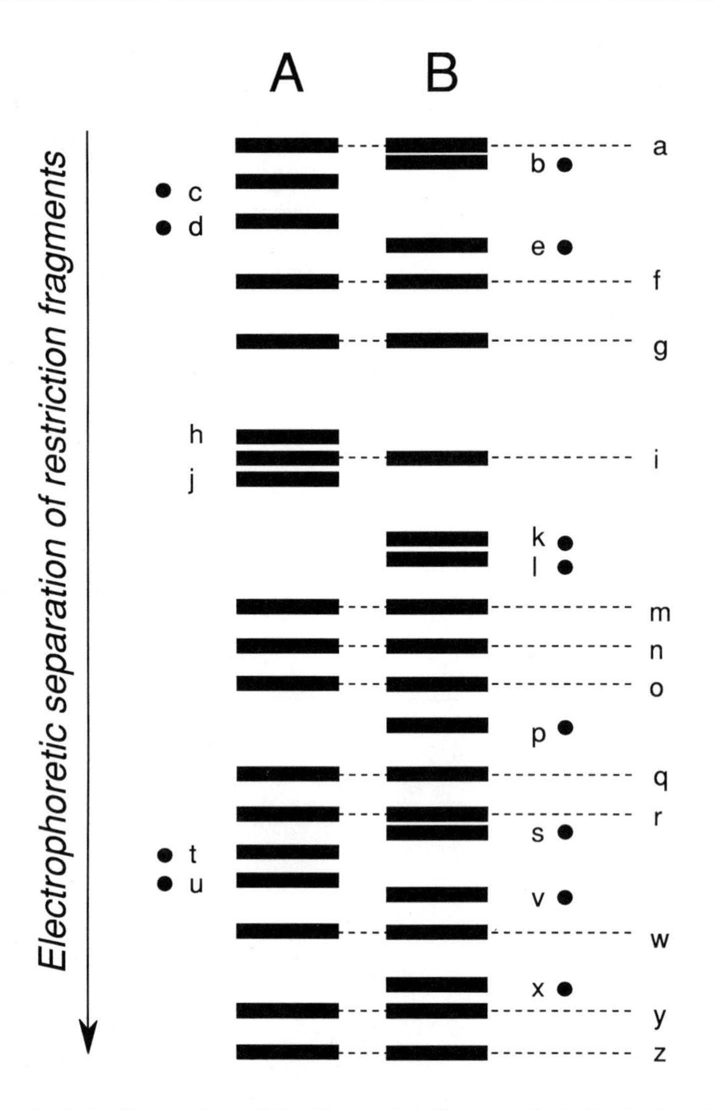

Figure 5. DNA fingerprints. DNA fingerprints from two individuals, A and B, composed of electrophoretically separated DNA fragments that appear as bands. Shared bands are indicated by dashed lines connecting the bands and their identifying letter designation. Bands unique to the individuals are denoted by dots along side of the letter designations.

dicates relatedness or chance convergence of fragment sizes. Thus, multiple-site VNTRs cannot be used to determine degrees of relatedness beyond parent-offspring, and cannot be used to measure the level of inbreeding (Lynch 1989). Specific or single-site VNTRs do exist, however, and these polymorphic loci may be useful for linkage studies if their inheritance can be determined (Nakamura et al. 1987).

Molecular cloning/PCR

A new method for rapid enzymatic amplification of specific DNA sequences, called the "polymerase chain reaction" or PCR (Saiki et al. 1988a), has obviated the need for recombinant DNA cloning methods in many cases and has opened up the possibility of directly obtaining nucleotide sequence information for individuals (Gyllensten and Erlich 1988). Visualization of differences at the level of DNA sequence is the most unambiguous and precise method for detecting genetic variation. Indeed, individuals that might appear to be identical by allozyme or RFLP tests may be distinguishable at the DNA-sequence level. Because of the significance of PCR, we review its features briefly (see Saiki et al. 1988b for details). PCR is being used to amplify both mitochondrial and nuclear genes.

Polymerases — enzymes responsible for DNA replication — synthesize the second-strand complement of single-stranded DNA (ssDNA), beginning at the point where the ssDNA extends from a double-stranded DNA (dsDNA) molecule. A polymerase can therefore be directed to copy a particular DNA segment by allowing oligonucleotides complementary to sequences flanking the region of interest to anneal to ssDNA obtained by melting native dsDNA. The polymerase will then specifically synthesize copies of the gene of interest. The ingredients of a PCR are genomic DNA (i.e., all of the DNA that may be isolated by digestion of tissues and chemical extraction), oligonucleotide primers, free de-oxyribonucleotides (the building blocks from which the DNA is constructed), and a DNA polymerase called "*Taq* polymerase" or simply "*Taq*" from *Thermus aquaticus*, a bacterium that normally lives at very high temperatures.

Figure 6 illustrates a typical amplification of a specific DNA sequence by means of the PCR technique. The process begins with thermal denaturation of genomic double-stranded DNA (dsDNA) which melts into

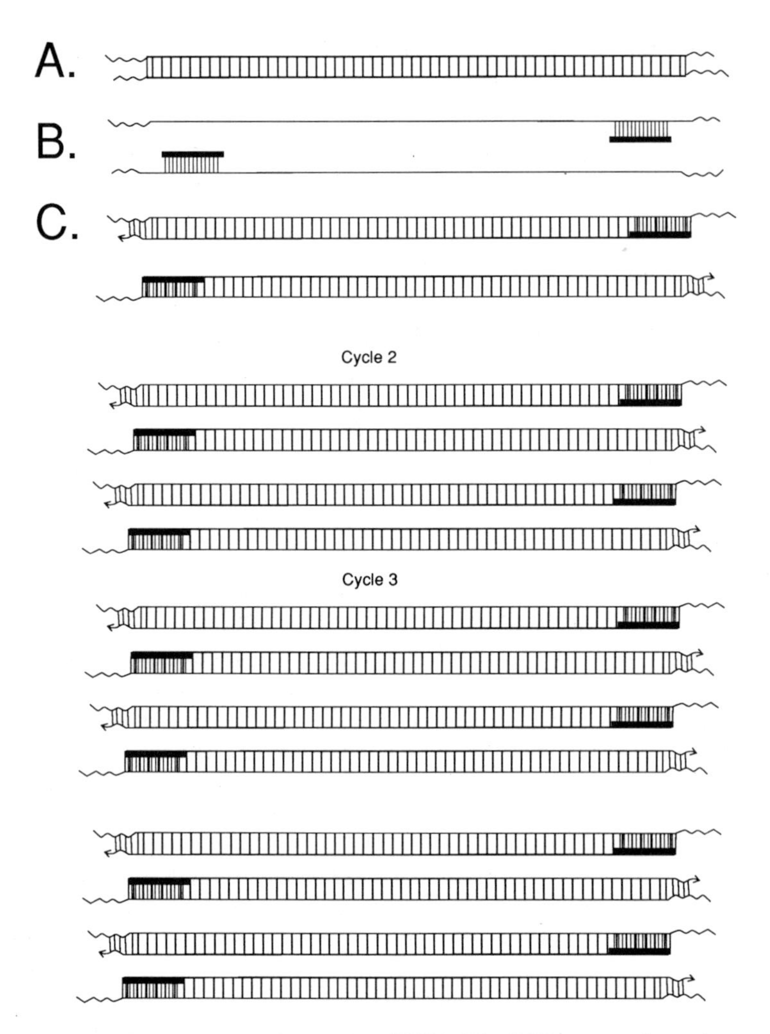

Figure 6. Polymerase chain reaction (PCR). The PCR process is used to greatly amplify (i.e., produce large quantities of) specific gene sequences. A: Cross-hatched area indicates targeted sequences of DNA. B: Double-stranded DNA is heated to separate complementary strands. Single-stranded DNA primers (dark bars) are allowed to anneal to complementary sequences flanking the targeted DNA. C: The polymerase initiates a synthesis ("polymerization") of new complementary DNA strands. This cyclic series of events is repeated, resulting in the exponential accumulation of the target DNA sequence.

single-stranded DNA (ssDNA) in a minute or less at 94 C. The reaction is then cooled to 45-55 C for a minute or less to allow for the annealing of the primers to the template. Finally, extension of the new strands proceeds at 72 C for 1 minute per 1,000 bases. Because *Taq* is naturally adapted to high temperatures, this enzyme, unlike polymerases from other sources, survives the DNA denaturation step. Use of *Taq* polymerase allows automation of PCR, and programmable heat blocks are commercially available for this purpose. With PCR, accumulation of copies of the target DNA is theoretically exponential, so that 10^8-fold amplification is possible in about 3 hours. If the primers are in asymmetric ratios (1:50 or 1:100) the reaction synthesizes an excess of single-stranded DNA which can then be sequenced directly using commercially available DNA-sequencing kits (Gyllensten and Erlich 1988).

There are several advantages of PCR. First, it is both sensitive and specific. In principle, as long as one target molecule is present in a sample, it can be amplified (i.e., you can pull the genetic needle out of the genomic haystack). Second, elaborate procedures for DNA isolation are frequently not necessary because the target is amplified away from contaminants. Third, PCR can be used for very small samples: a single larva, tissue from a single walking leg, bacterial samples, etc. Fourth, PCR allows use of non-radioactive labeling methods because the concentration of the target can be increased to the point where these less sensitive detection methods work.

Sequence-specific typing/ASO and SSO probes

Oligonucleotides that correspond to particular alleles or DNA sequences in a population ("allele-specific", or "sequence-specific oligonucleotides," ASOs or SSOs respectively), can be synthesized and used to type new unknown samples without need of either complete sequencing or radioactive labeling (Saiki et al. 1988c). As little as a single nucleotide base-pair difference between alleles can be detected, owing to the greater stability of perfectly matched vs. imperfectly matched oligonucleotide hybridization at a given temperature and salt concentration. Figure 7 illustrates the complementary and non-complementary hybridization between an ASO probe and two different allelic sequences. Figure 8 illustrates the procedure for typing of test samples. Samples of PCR-amplified DNA are immobilized on a nylon membrane solid support. The membrane is then

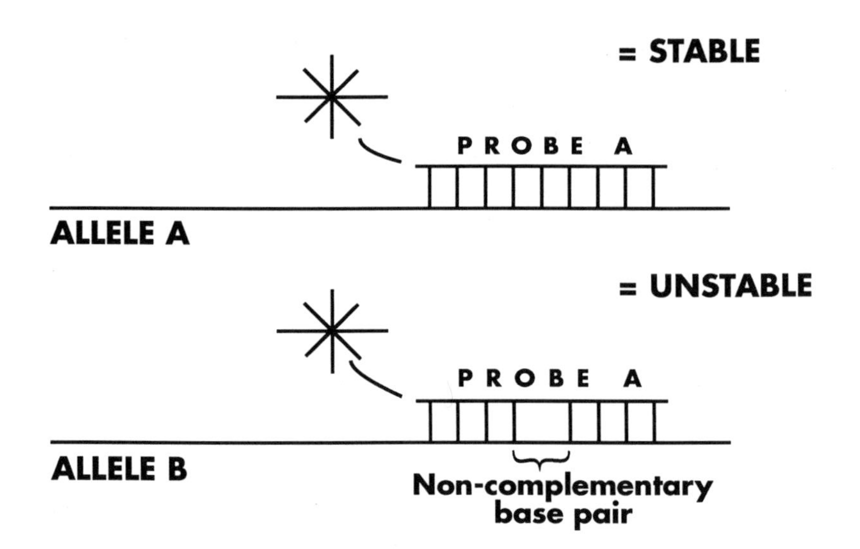

Figure 7. ASO binding. Binding of an ASO (allele-specific oligonucleotide) probe to completely complementary (allele A) and incompletely complementary (allele B) nucleotide base sequences. In the case of allele A, the base-pair sequence of the probe exactly matches the base sequence of the allele and probe hybridization is stable. In the case of allele B, the binding of the probe is destabilized by the single base-pair difference where no binding of probe and allele occurs.

soaked in a solution containing an ASO probe which has been conjugated to the enzyme horseradish peroxidase (HRP). If the ASO matches the sample and sticks, its presence is detected by a color-producing reaction catalyzed by the HRP. The membrane can be washed and retested with additional ASOs until a match is found. An alternative testing format, in which the ASOs are immobilized on a membrane strip and HRP-binding PCR product is applied to the strip containing all allelic possibilities, has been developed for prenatal diagnosis and forensic typing in humans (Bugawan et al. 1988; Saiki et al. 1989). These methods could easily be adapted to pedigree analysis or strain identification in aquaculture brood stocks.

Marker-assisted selection

The generation of numerous genetic markers by the application of molecular methods such as those described above enables aquaculture

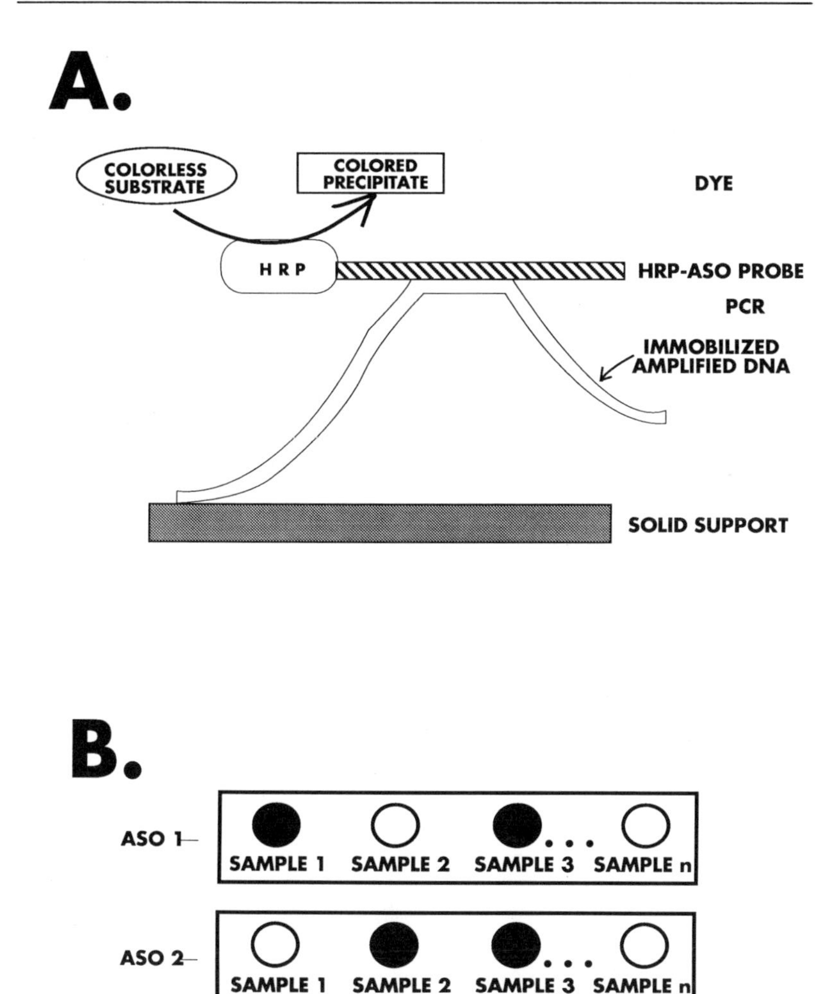

Figure 8. ASO typing procedure. A. PCR-amplified DNA sequences have been immobilized on a solid support. Horseradish peroxidase (HRP) is conjugated to an ASO probe for specific DNA sequences in a sample. The HRP-ASO probe is hybridized with the immobilized sample. If the ASO probe sequence matches the sample sequence, the probe will stably bind to the sample (as in Fig. 7). HRP catalyzes a reaction which turns a colorless enzyme substrate to a colored precipitate. B: Sample DNA sequences giving positive, (dark circles) and negative (open circles) reactions to two (1,2) ASO probes. A positive reaction means the sample contains the exact sequence of the ASO probe.

breeders to look for correlations among the markers and genes controlling quantitative traits ("quantitative trait loci" or QTLs). Examples of quantitative traits controlled by QTLs in prawns and shrimp could be growth rate, food conversion efficiency, disease resistance, and body proportions. If QTL-genetic marker linkages exist, then selection can be more efficiently focused on the markers rather than on the quantitative traits (Soller and Beckmann 1988). This type of selection is called "marker-assisted selection" or MAS. The advantage of MAS is illustrated in Figure 9. As shown in the figure, the response to selection (increase of genetic value over gen-

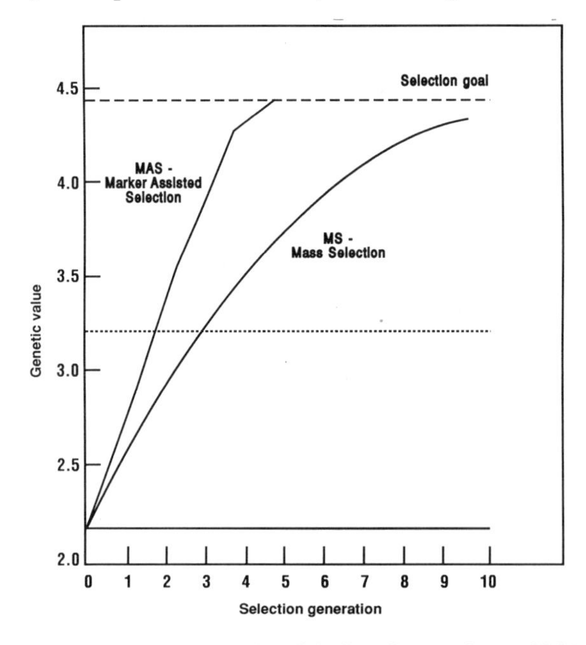

Figure 9. Marker assisted selection (MAS). Comparison of MAS and Mass Selection (MS) in a hypothetical, synthetic population obtained by crossing individuals in a commercial stock. In MS, selection is based upon the individual's phenotype controlled by quantitative trait loci (QTLs). In MAS, the selection is based upon the presence of linked genetic markers. The two curves shown represent the increase in a hypothetical genetic value (y axis) over time in generations (x axis). The curves were derived from a model developed by Soller and Beckmann (1988). The dotted line represents the breeding goal. Clearly MAS achieves the goal in four generations, while MS requires more than twice this amount of time. (Redrawn from Soller and Beckmann (1988) with permission).

erations) is much faster in MAS than in mass selection (MS) based on the trait controlled by QTLs. Inbred lines that have different marker alleles and that differ dramatically in quantitative trait expression have been used in crops and poultry to find QTL-marker linkages (Edwards et al. 1987; Soller and Beckmann 1988; Stuber 1988). Unfortunately, no inbred lines of shrimp or prawns are yet available for similar studies.

Another situation in which genetic markers can be used is in crossing a commercial stock whose present value is well known, with a resource population — perhaps a wild stock which has different marker alleles and which differs in some desirable trait such as disease resistance (Soller and Beckmann 1988). Without very strong selection, the frequency of a favorable allele that the breeder is trying to incorporate into the commercial stock through backcrossing is rapidly lost through segregation. With MAS however, an allele linked to a marker can be maintained at its maximum frequency of 0.5. This is illustrated in Figure 10, where the frequency of a favorable allele is shown to be maintained at the maximum frequency level in a MAS stock as compared to MS stocks. It must be noted that the application of MAS assumes that a commercial stock exists, which is not the case in shrimp and prawn aquaculture.

Soller and Beckmann (1988) estimate that the cost of RFLP-based QTL analysis might run as high as $0.5 to 1 million dollars, but that economic gain can offset this cost. For example, these authors show that the introduction of a gene whose effect is to increase hen laying capacity by 6 eggs per year would have an impact in the tens of millions of dollars in the egg industry. Presently, high producing hen flocks can lay over 200 eggs per hen per year (Nordskog 1977). Economic analyses of MAS should be done for shrimp and prawn aquaculture, although there are presently few bioeconomic models available to conduct these types of sensitivity analyses (Malecha and Hedgecock 1989). Finally, cost per sample using DNA-based marker technologies may drop as these methods become more fully automated (Landegren et al. 1988).

Manipulation of Reproductive Processes

Just as the manipulation of reproduction in farm animals has revolutionized breeding methods (Womack 1987; Smith 1988), control over re-

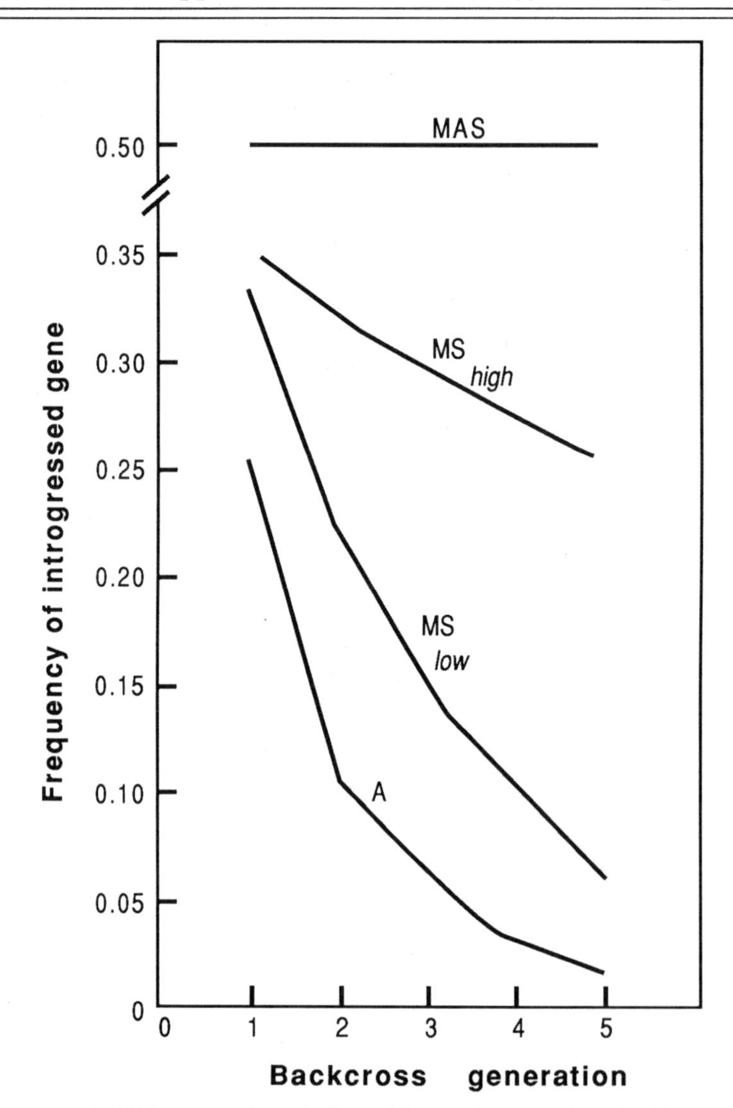

Figure 10. MAS incorporation of a favorable gene into brood stock. Comparison of the frequency of a favorable allele that a breeder is trying to incorporate (i.e., "introgress") and maintain in a commercial stock through backcrossing when selection is based upon MAS and MS under two levels of selection intensity, high and low. The curve labelled A refers to the frequency of the allele in an unselected population. (Redrawn from Soller and Beckmann (1988) with permission).

production in aquaculture (either through hormonal or genetic manipulation of sex determination, or through manipulation of the number of sets of chromosomes) has been an important area of research and commercial application (Purdom 1983; Yamazaki 1983; Allen 1987). Because these developments have been reviewed thoroughly for fish and mollusks (op. cit.), we will briefly discuss potential applications of these technologies to shrimp and prawn breeding.

Cryopreservation of sperm, eggs and embryos

Cryopreservation of sperm is commonly used in breeding domesticated animals through artificial insemination (AI). A chief use of AI technology for shrimp and prawns would be to carry out the large scale breeding experiments needed to determine the genetic basis of variation in quantitative traits (Malecha and Hedgecock 1989). These mating experiments require simultaneous matings, which are difficult to achieve given the present technology of brood stock management. The sperm of a marine shrimp have been successfully cryopreserved (Anchordoguy et al. 1988), so there is little doubt that this technology could be applied to shrimp breeding experiments. Cryopreservation of eggs and embryos would provide an added measure of control in mating experiments (A. Lawrence, personal communication).

Chromosome set manipulations — polyploidy

Female marine shrimp spawn eggs that are arrested in metaphase of meiosis I (Clark et al. 1984; Pillai et al. 1988). If the female has been mated, the eggs are fertilized as they pass over the sperm storage receptacles. Blocking either of the two subsequent maturational divisions, or even the first cleavage division of these eggs, would have dramatic genetic and possibly physiological consequences. Polyploids are expected to result, either triploids (two kinds), or tetraploids, depending respectively on whether a maturational or a cleavage division is inhibited (Fig. 11). Physical or chemical treatments have been used to block egg divisions in fish and mollusks, and the resulting triploids have been commercially useful (Purdom 1983; Allen 1987). To our knowledge, methods for inducing polyploidy in prawns and marine shrimp have not been explored, and appropriate treatments for arresting egg divisions have not yet been determined.

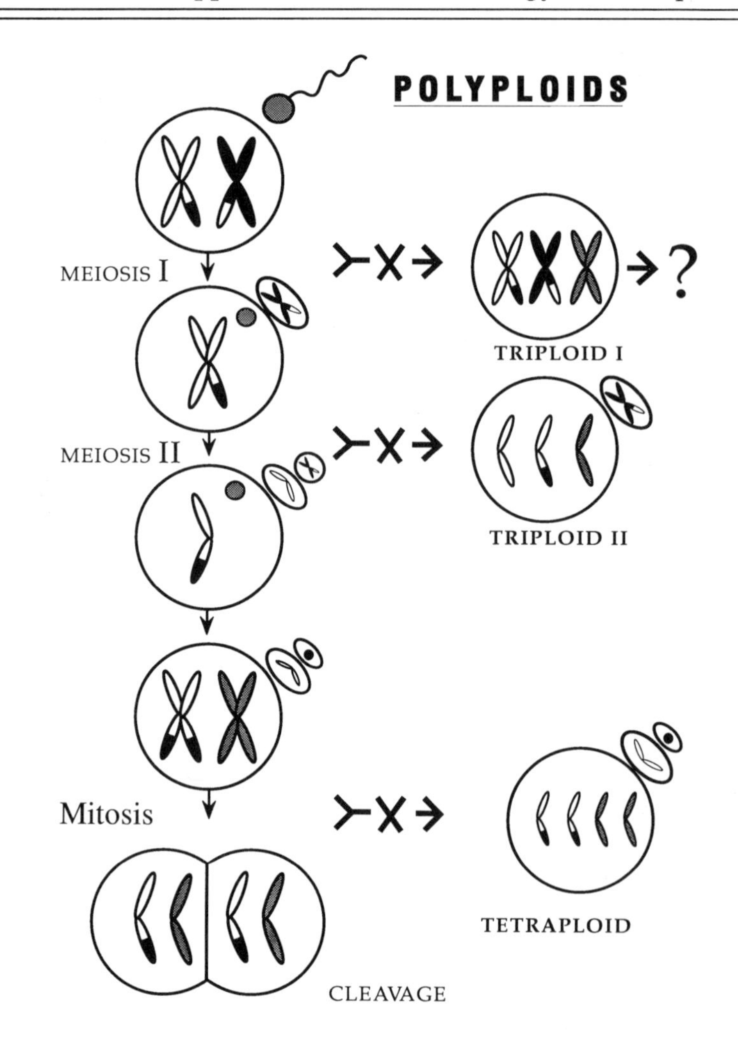

Figure 11. Induction of polyploidy. Chromosomal configurations during oogenesis (meiosis) showing how triploid and tetraploid offspring may be produced. The cell representations on the left show the normal sequence of cellular division during meiosis and first mitosis. The fate of only one chromosomal pair is portrayed for the sake of simplification; polyploidization can occur for all of the chromosomes in the oocyte. Interventions (X) at either of the meiotic divisions may result in a triploid zygote having two sets of maternal and one set of paternal chromosomes; blockage of first mitosis could result in a tetraploid.

Chromosome set manipulations — artificial parthenogenesis

Parthenogenesis — the process of creating diploid progeny without a paternal chromosomal contribution — is shown in Figure 12. In marine shrimp, eggs can be activated to undergo meiosis upon contact with seawater (Clark et al. 1984; Pillai et al. 1987), and in the absence of sperm, the egg proceeds through meiosis, thereby becoming haploid (i.e., possessing only one set of maternal chromosomes). A haploid embryo undergoes abnormal cell division and dies in the early cleavage stages. Blocking one of the egg-maturational divisions or the first cleavage could, in principle, restore diploidy and produce a viable but inbred zygote. This method might be used to produce inbred lines of prawns and marine shrimp more rapidly than by the brother-sister matings practiced in poultry breeding. Inbred lines would, of course, be extremely useful in genetic research; however, they might also be important in crossbreeding programs for stock improvement. We know of no research efforts for either prawn or shrimp aquaculture in this area.

Sex-determination and sex-ratio control

Sexual dimorphism in production characteristics for prawns implies that production could be improved through control of sex ratio in cultured populations. In particular, the growth-suppressing effects of dominant males in mixed-sex prawn culture could be avoided (Ra'anan and Cohen 1985; Sagi et al. 1986). Malecha et al. (1990) obtained complete sex-reversal of female prawns to functional males by implanting androgenic gland tissue in very small, sexually undifferentiated females. The sex-reversed "neomales" (genetic females) developed complete reproductive competence and were able to produce progenies with skewed sex ratios.

The sex determination mechanisms of prawns were partially elucidated through the use of these neomales in matings with normal females, and the intercrossing and backcrossing of the F1 progeny of such crosses (Malecha et al. 1990). The presence of males in these progenies rules out a chromosomal sex determination in which the female is homogametic (XX), as in humans. Sex ratios skewed towards females instead suggest a chromosomal sex determination mechanism in which males are homogametic (ZZ) and females are most often heterogametic (ZW) but possibly also ho-

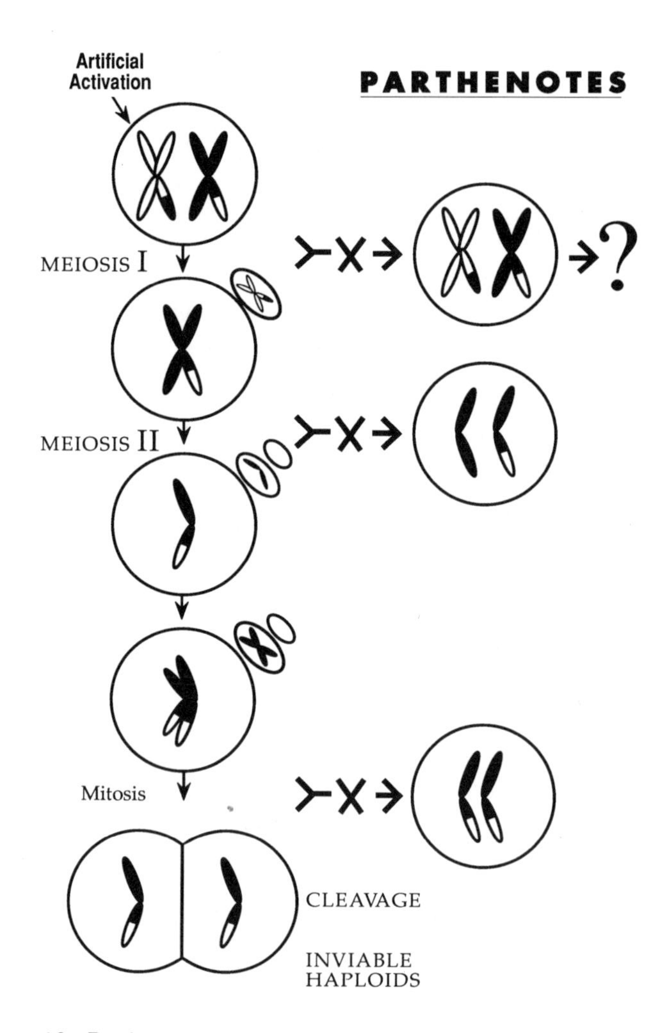

Figure 12. Parthenogenesis. Chromosomal configuration during oogenesis showing how parthenotes can be produced by artificial activation followed by blocking of a cell division. Chromosomes are portrayed as described in Figure 11. In parthenogensis, only a maternal chromosome complement is involved so the resulting zygotes are inbred to varying degrees depending on which cell division is interrupted.

mogametic (WW). Great variation in the sex ratios of individual families suggests, however, that control of sex must be even more complex than this simple chromosomal mechanism. Whatever the final explanation, these experiments demonstrate the ability to alter sex ratios, and also point the way toward eventual control over the sex of cultured prawns.

Finally, from a heterogametic female sex-determining system it follows that ablation of the androgenic gland using modifications of the procedures described by Nagamine et al. (1980b) should reverse genetic males (ZZ) to phenotypic females ("neofemales"). Mating "neofemales" to normal males should yield all males (ZZ) in the F1 generation. If commercial-level technology for producing either all-male or predominantly male as well as all-female or predominately female progeny were available, then aquaculturists would have powerful economic options for maximizing commercial production.

Knowledge of both genetic and molecular bases of sex determination in other organisms (like the nematode *Caenorhabditis elegans* and the fruit fly *Drosophila*) is accumulating rapidly (Nöthiger and Steinmann-Zwicky 1985; Hodgkin 1987). As individual genes involved in sex determination are identified and their modes of action understood, it should be possible to determine whether similar sex-determining genes are present in shrimp and prawns. One approach might be to use oligonucleotide primers corresponding to conserved regions either within or flanking those genes for PCR amplifications of these genes (should they exist) in shrimp DNA. Such an approach has been taken in the study of mtDNA evolution; "universal" primers have been developed that direct PCR amplification of homologous mtDNA sequences from a wide taxonomic array (Kocher et al. 1989). If sex-determining genes were successfully amplified from shrimp and prawns, then the next steps would be to examine their expression during development, and then to determine the nature of the genetic regulation of that expression. Knowing the DNA sequence of these genes would enable the synthesis of oligonucleotide probes that could be used to screen for gene expression in various tissues during development (via hybridization with messenger RNA or its DNA equivalent). Amplification of control regions surrounding the target sex-determining genes could be accomplished by means of an inverse PCR strategy (Ochman et al. 1988). Clearly, in the near future, advances in molecular biology that are made using certain experimental organisms may be rapidly applied to other organisms (such as shrimp and prawns) that are not directly used in experimental molecular biology.

Ancillary Uses of Markers or Marker Technology

The notion that the transfer of biotechnology to species other than those particularly suited to laboratory study may be accelerated by new molecular tools (such as PCR and the automation of molecular cloning and sequencing methods) implies an impending revolution in the next decade in the way that biological problems are solved and processes understood. While it is impossible for us to comment on the potential for advances in all fields of shrimp and prawn biology, we will mention briefly two areas in which advances might have considerable impact on future breeding programs: the physiology of crustacean growth and reproduction and microbial identification, including pathogen detection.

The physiology and endocrine regulation of crustacean growth and reproduction is beginning to be understood in considerable detail (Chang 1989; Dougherty 1989). In the same way that we anticipate advances in knowledge of sex determination to be transferable to shrimp and prawns, we anticipate that advances in identification of hormones controlling growth or reproduction, and the genes encoding them or regulating their expression, will spread rapidly. Again, tools such as PCR and SSO hybridization will likely play an important role in locating homologous genes, in detecting patterns of gene expression and in identifying sequences involved in regulation of expression.

Recent publication of the use of fluorescence-labeled SSOs for the microscopic identification of single microbial cells promises a revolution in the studies of microbial ecology and pathology (DeLong et al. 1989). The method is based on the use of oligonucleotides complementary to taxa-specific regions of ribosomal RNA, which is abundant in microbial cells. One probe distinguishes among the eukaryotes, the eubacteria and the archeobacteria; other probes can distinguish among much more closely related microbes. Combinations of probes labeled with different fluorescent dyes can be employed simultaneously. Probe hybridization is done on formaldehyde-fixed intact cells on microscope slides, and the results viewed under a fluorescence light microscope. The potential for employing this simple method in studying the microbial ecology of hatcheries and ponds, and in pathogen detection and disease diagnosis seems large.

We also envision the possibility of using PCR to develop probes to diagnose the presence of IHHN or baculoviruses in shrimp brood stock. IHHN is a lethal or semi-lethal viral disease (Lightner 1988) whose only "cure" is prevention. If aquaculturists had a means for early diagnosis of IHHN, they could maintain IHHN-free brood stock.

ADVANCED BIOTECHNOLOGY - DIRECT GENOME INTERVENTION

Visual images associated with some of the more spectacular achievements of genetic engineering — transgenic mice the size of rats (Palmiter et al. 1982; Palmiter et al. 1983) — understandably capture the public's attention and imagination. We fear, however, that such images create unrealistic expectations for the genetic improvement of aquaculture species. We believe that contributions to shrimp and prawn aquaculture from genetic engineering defined as direct genomic intervention are far in the future, particularly when one examines critically the state of biological knowledge and brood stock management technology for these species. Nevertheless, methods of molecular biology will contribute substantially to the development of appropriate, fundamental husbandry and breeding practices.

Lest the reader think us overly pessimistic about the prospects for genetic engineering of shrimp and prawns, we point to Schuler's (1988) recent remark concerning genetic engineering in agriculture: "Strictly speaking, I know of no clear-cut case of an introduction of a single gene altering a *quantitative* trait." The general consensus among animal scientists concerned with the quantitative traits underlying production characteristics, is that genetic engineering may help but will never replace traditional methods of animal improvement. Smith (1988) provides an excellent discussion of the economic constraints on direct genome intervention in animal breeding.

Here, we will review briefly some of the major constraints on the application of direct genomic intervention in shrimp and prawns. Basic research on shrimp and prawn biology may remove some, but probably not all, of these constraints in the coming decade.

Microinjection vs. Vector-Mediated Insertion

Direct manipulation of the genomes of shrimp and prawns, whether by means of the directed mutation of an endogenous gene or the introduction of an exogenous gene (a "transgene"), will depend on the development of either some means of physically introducing DNA into eggs (such as microinjection), or the discovery of viruses, retroviruses or transposable elements that may serve as vectors for transferring exogenous genes into the genomes of shrimp or prawns. Microinjection has been used both in mammals (Gordon and Ruddle 1981; Palmiter et al. 1982; Palmiter et al. 1983; Hammer et al. 1985) and fish (Maclean et al. 1987). However, with microinjection, there is little or no control over the number of copies of the piece of DNA introduced, its ultimate location in the genome (should it even integrate into the host genome), or the physical damage resulting from such an invasive technique. Retroviruses or transposable elements, on the other hand, have advantages over microinjection (Crittenden and Salter 1988), but have not yet been found in decapod crustaceans.

Discussion of the technical problems of gene insertion presupposes that desirable mutations or foreign genes can be identified, and that stocks exist whose breeding value is known sufficiently to allow some estimate of the expected gain from attempting direct genome intervention; neither prerequisite exists for shrimp or prawns.

Directed Mutagenesis

Technology exists for introducing specific changes into particular DNA sequences. A mutant oligonucleotide primer that differs from the "wild type" DNA sequence by a single base can be amplified using PCR to amounts sufficient for microinjection or vector-mediated transduction experiments (Higuchi et al. 1988). The problem is then one of identifying which mutations are desirable. Advances in protein chemistry are certainly bringing us closer to the era of "designer enzymes" and may someday allow us to make amino acid modifications to particular enzymes that might enhance production. But it appears that we are a very long way from even identifying particular gene products that have such major effects on shrimp and prawn production, much less from changing them through directed mutagenesis.

Transgenic Incorporation

Smith (1988) distinguishes between transgenes having large effects (e.g., disease resistance, growth hormones), and those having smaller effects on economic merit in conventional production systems. In either case, the basic genetics of transgene introduction are the same. Figure 13 illustrates the procedure of introducing an exogenous gene (transgene, T) into a commercial stock. The original transgenic founder will likely be hemizygous (T/0) for the foreign gene, compared to normal individuals lacking the transgene (0/0). It will likely be most desirable to have individuals homozygous (T/T) for the transgene, and this must be done by first mating transgenic individuals to commercial brood stock, and then backcrossing the F_1 offspring to the T/0 parent. The first task is to evaluate the fitness of the T/T homozygote, and then the next step would be to incorporate the transgene into the commercial stock by either repeated backcrossing or introgression. Again, MAS would make this introgression more efficient, so it may prove useful to incorporate PCR and SSO typing of the transgene construct in order to follow its inheritance. Each transgenic individual must undergo these two steps for evaluation and incorporation because the site of transgene integration into the host genome will likely be uncontrolled and therefore unique to each.

Clearly, the steps needed to evaluate and incorporate foreign DNA into commercial stocks require resources and facilities beyond those required for traditional brood stock development. Recall that facilities for traditional brood stock development do not yet exist for shrimp and prawns (Malecha and Hedgecock 1989). Unless a high value stock is utilized in the first place, the economic benefit of all the backcrossing (with its attendant time lag) would be very doubtful. Smith (1988) estimates that, for the introduction to be worthwhile, a transgene should increase economic value by at

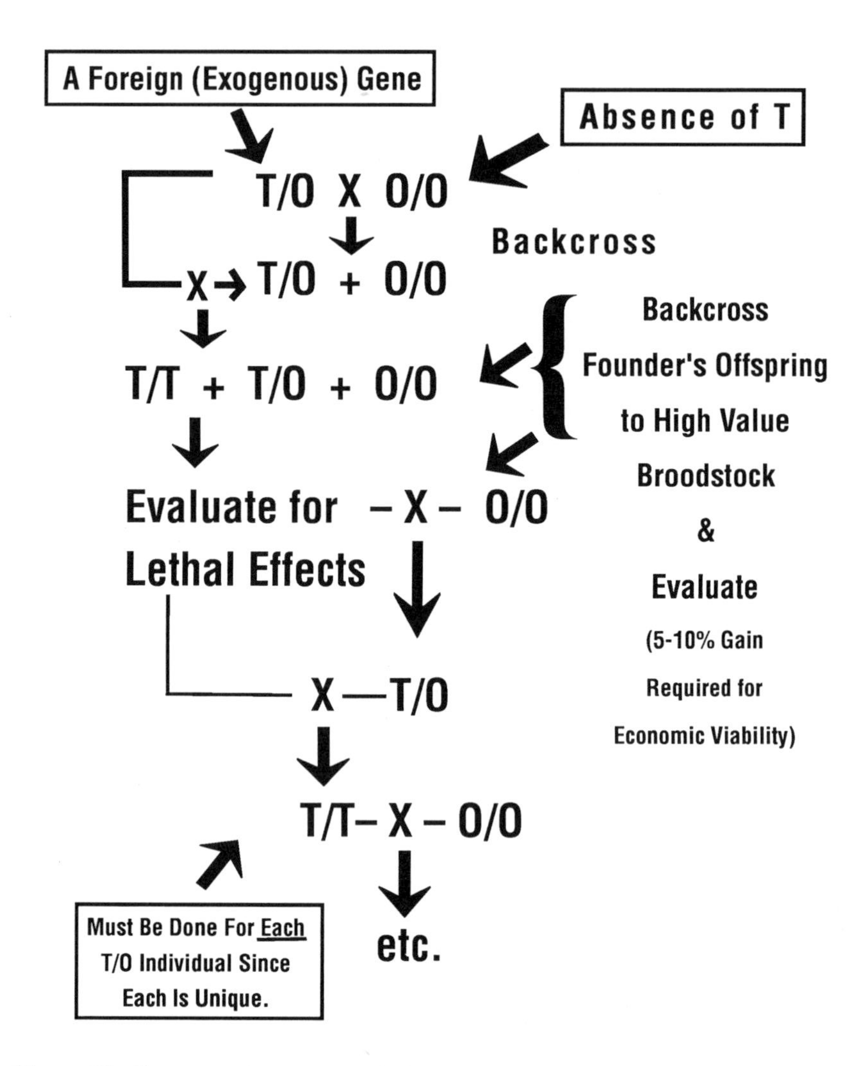

Figure 13. Incorporating transgenes into commercial stock. An individual hemizygous, (T/O) for foreign genes is first mated with a homozygous (O/O) individual. This is followed by a backcross of the T/O progeny to the T/O parent to produce homozygous T/T individuals who are then evaluated for lethal or other detrimental effects. This is followed by backcrossing and introgression to create a completely homozygous (T/T) population.

least 5 -10%. Shrimp and prawn aquaculturists do not yet have pedigreed commercial stocks nor the means, therefore, of evaluating the relative economic merit of different stocks (Malecha and Hedgecock 1989).

CONCLUSIONS

It is very unlikely that genetic engineering by direct genomic intervention and modification will contribute to shrimp and prawn aquaculture in the next decade. Basic research is needed on methods for inserting DNA into the germ lines of these crustaceans, and on identifying particular genes that affect production characteristics.

It is far more likely that biotechnology will be employed in concert with the development of husbandry and breeding methods that are required for domestication and improvement of shrimp and prawn brood stocks. Biotechnology is likely to be directly employed in breeding programs in three areas: (1) genetic markers, which are important in identifying individuals, keeping pedigrees, and in assisting selection through linkage to genes affecting quantitative (production) traits; (2) manipulations of gametes (cryopreservation), and chromosome sets (induced polyploidy or artificial parthenogenesis); and (3) sex control, breeding of sex-reversed brood stock for mono-sex or sex ratio-controlled culture. Finally, biotechnology is likely to affect brood stock development programs indirectly through advances in microbial ecology, pathology and in basic molecular, cellular and organismal biology.

LITERATURE CITED

Allen, S. K., Jr. 1987. Genetic manipulations — critical review of methods and performances for shellfish. Pages 127-144 *in* K. Tiews, editor. Selection, hybridization and genetic engineering in aquaculture. Heenemann Verlag, Berlin .

Anchordoguy, T., J.H. Crowe, F.J. Griffin, and W.H. Clark Jr. 1988. Cryopreservation of sperm from the marine shrimp *Sicyonia ingentis*. Cryobiology 25:238-243.

Avise, J. C. 1986. Mitochondrial DNA and the evolutionary genetics of higher animals. Philosophical Transactions of the Royal Society, London B312:325-342.

Brown, W. M. 1983. Evolution of animal mitochondrial DNA. Pages 62-88 *in* M, Nei, and R.K. Koehn, editors. Evolution of Genes and Proteins. Sinauer Assoc., Sunderland, Massachusetts, pp. 62-88.

Bugawan, T. D., R.K. Saiki, C.H. Levenson, R.M. Watson, and H.A. Erlich. 1988. The use of non-radioactive oligonucleotide probes to analyze enzymaticaly amplified DNA for prenatal diagnosis and forensic HLA typing. Biotechnology 6:943-947.

Chang, E. S. 1989. Endocrine regulation of molting in crustaceans. Reviews in Aquatic Science 1:131-157.

Charniaux-Cotton, H. and G. Payen. 1988. Crustacean reproduction. Pages 279-303 *in* H. Laufer and R. G. H. Downer, editors. Endocrinology of selected invertebrate types. Alan R. Liss, Inc., New York.

Clark, W. H., Jr., A.I. Yudin, F.J. Griffin, and K. Shigekawa. 1984. The control of gamete activation and fertilization in the marine Penaeidae, *Sicyonia ingentis*. Pages 459-472 *in* W. Engels, W. H. Clark, Jr., A. Fisher, P. J. W. Olive, and D. F. Went, editors. Advances in invertebrate reproduction, Vol 3., Elsevier, New York.

Crittenden, L. B. and D.W. Salter. 1988. Insertion of retroviruses into the avian germ line. Pages 207-214 *in* B. S. Weir, E. J. Eisen, M. M. Goodman, and G. Namkoong, editors. Proceedings of the Second International Conference on Quantitative Genetics. Sinauer Assoc., Sunderland, Massachusetts.

DeLong, E. F., G.S. Wickham, and N.R. Pace. 1989. Phylogenetic stains: ribosomal RNA-based probes for identification of single cells. Science 243:1360-1363.

Dougherty, W., editor. 1989. Frontiers of shrimp research. Elsevier, New York, in press.

Edwards, M. D., C.W. Stuber, and J.F. Wendel. 1987. Molecular-marker-facilitated investigations of quantitative-trait loci in maize. I. Numbers, genomic distribution and types of gene action. Genetics 116:113-125.

Ferris, S. D. and W.J. Berg. 1987. The utility of mitochondrial DNA in fish genetics and fishery management. Pages 277-299 *in* N. Ryman and F. Utter editors. Population genetics and fishery management. University of Washington Press, Seattle.

Gordon, J. W. and F.H. Ruddle. 1981. Integration and stable germ line transmission of genes injected into mouse pronuclei. Science 214:1244-1246.

Gyllensten, U. and A.C. Wilson. 1987. Mitochondrial DNA of salmonids: inter- and intraspecific variability detected with restriction enzymes. Pages 301-317 *in* N. Ryman and F. Utter editors. Population genetics and fishery management. University of Washington Press, Seattle, WA.

Gyllensten, U. B. and H.A. Erlich. 1988. Generation of single-stranded DNA by the polymerase chain reaction and its application to direct sequencing of the HLA-DQA locus. Proceedings of the National Academy of Sciences USA 85:7652-7656.

Hammer, R. E., V.G. Pursel, C.E. Rexroad, Jr., R.J. Wall, D.J. Bolt, K.M. Ebert, R.D. Palmiter, and R.L. Brinster. 1985. Production of transgenic rabbits, sheep and pigs by microinjection. Nature 315:680-683.

Hedgecock, D. 1977. Biochemical genetic markers for broodstock identification in aquaculture. Proceedings of the World Aquaculture Society 8:523-531.

Hedgecock, D., R.A. Shleser, and K. Nelson. 1976. Applications of biochemical genetics to aquaculture. Journal of the Fisheries Research Board of Canada 33:1108-1119.

Hedgecock, D., D.J. Stelmach, K. Nelson, M.E. Lindenfelser, and S.R. Malecha. 1979. Genetic divergence and biogeography of natural populations of *Macrobrachium rosenbergii*. Proceedings of the World Mariculture Society 10:873-879.

Hedgecock, D., K. Nelson, and M.L. Tracey. 1982. Genetics. Pages 283-403 *in* L. G. Abele, editors. The biology of Crustacea, Vol. 2. Academic Press, New York.

Hedgecock, D. and F. Sly. 1990. Genetic drift and effective population sizes of hatchery-propagated stocks of the Pacific oyster *Crassostrea gigas*. Aquaculture 88:21-38.

Higuchi, R., B. Krummel, and R.K. Saiki. 1988. A general method of in vitro preparation and specific mutagenesis of DNA fragments: study of protein and DNA interactions. Nucleic Acids Research 16:7351-7367.

Hodgkin, J. 1987. Sex determination and dosage compensation in *Caenorhabditis elegans*. Annual Review of Genetics 21:133-154.

Jeffries, A. J., V. Wilson, and S.L. Thein. 1985. Individual- specific "fingerprints" of human DNA. Nature 316:76-79.

Kawagashi, D. K., K.M. McGovern, D.L. Pavel, S.B. Ashmore, and N.P. Carpenter. 1986. Monitoring reproductive performance of *Penaeus vannamei* under commercial conditions. Paper presented at World Mariculture Society Annual Meeting, Reno, Nevada, January 1986.

Kocher, T. D., W.K. Thomas, A. Meyer, S.V. Edwards, S. Pbo, F.X. Villablanca, and A.C. Wilson. 1989. Dynamics of mitochondrial DNA evolution in animals: amplification and sequencing with conserved primers. Proceedings of the National Academy of Sciences USA 86:6196-6200.

Kulbarni, G. K., R. Nagabhushanan, and P.K. Joshi. 1984. Neuroendrocrine control of reproduction in the male penaeid prawn, *Parapenaopsis hardwickii* (Miers) (Crustacea, Decapoda, Penaeidae). Hydrobiologia 108(3):281-289.

Landegren, U., R. Kaiser, C.T. Caskey, and L. Hood. 1988. DNA diagnostics — molecular techniques and automation. Science 242:229-237.

Lester, L. J. 1979. Population genetics of penaeid shrimp from the Gulf of Mexico. Journal of Heredity 70:175-180.

Lester, L. J. 1983. Developing a selective breeding program for penaeid shrimp mariculture. Aquaculture 33:41-50.

Lightner, D. V. 1988. Diseases of cultured penaeid shrimp in the Americas. *in* C. J. Sinderman and D.V. Lightner, editors. Disease diagnosis and control in marine aquaculture in the Americas. Elsevier Scientific Publishing Co., Amsterdam, the Netherlands.

Lynch, M. 1988. Estimation of relatedness by DNA fingerprinting. Molecular Biology and Evolution 5:584-599.

Maclean, N., D. Penman, and Z. Zhu. 1987. Introduction of novel genes into fish. Bio/Technology 5:257-261.

Malecha, S. R. 1983. Crustacean genetics and breeding: an overview. Aquaculture 33:395-413.

Malecha, S. R. and D. Hedgecock. 1989. Prospects for domestication and breeding of marine shrimp. Technical Report UNIHI-SEAGRANT-TR-89-01, University of Hawaii Sea Grant College Program, Honolulu, Hawaii. Sept. 1989.

Malecha, S. R., P.A. Nevin, P. Ha-Tamaru, L.E. Barck, Y. Lamadrid-Rose, S. Masuno, and D. Hedgecock. 1990. Production of progeny from crosses of surgically sex-reversed freshwater prawns *Macrobrachium rosenbergii*: Implications for commercial culture of monosex populations. (Submitted).

Marx, J. L. 1988. Multiplying genes by leaps and bounds. Science 240:1408-1410.

Moav, R., T. Brody, G. Wohlfarth, and G. Hulata. 1978. Applications of electrophoretic genetic markers to fish breeding. I. Advantages and methods. Aquaculture 9:217-228.

Mulley, J. C. and B.D.H. Latter. 1981a. Geographic differentiation of eastern Australian penaeid prawn populations. Australian Journal of Marine and Freshwater Ressearch 32:889-895.

Mulley, J. C. and B.D.H. Latter. 1981b. Geographic differentiation of tropical Australian penaeid prawn populations. Australian Journal of Marine and Freshwater Research 32:897-906.

Nagabhushanan, R. and G.K. Kulbarni. 1981. Effects of exogenous testosterone on the androgenic gland and testes of a marine penaeid prawn, *Parapenaeopsis hardwickii* (Miers) (Crustacea, Decapoda, Penaeidae). Aquaculture 23:19-27.

Nakamura, Y., M. Leppert, P. O'Connell, R. Wolff, T. Holm, M. Culver, C. Martin, E. Fujimoto, M. Hoff, E. Kumlin, and R. White. 1987. Variable number of tandem repeat (VNTR) markers for human genetic mapping. Science 234:1616-1622.

Nelson, K. and D. Hedgecock. 1980. Enzyme polymorphism and adaptive strategy in the decapod Crustacea. American Naturalist 116:238-280.

Nordskog, A. W. 1977. Success and failure of quantitative genetic theory in poultry. Pages 569-586 *in* E. Pollack, O. Kenthorne, T. B. Bailey, editors. Proceedings international conference on quantitative genetics, August 1976. Iowa State University Press, Ames, Iowa.

Nöthiger, R. and M. Steinmann-Zwicky. 1985. Sex determination in *Drosophila.* Trends in Genetics, July: 209-215.

Ochmann, H., A.S. Gerber, and D.L. Hartl. 1988. Genetic applications of an inverse polymerase chain reaction. Genetics 120:621-623.

Palmiter, R. D., R.L. Brinster, R.E. Hammer, M.E. Trumbauer, M.G. Rosenfeld, N.C. Birnberg, and R.M. Evans. 1982. Dramatic growth of mice that develop from eggs microinjected with metallothionein-growth hormone fusion genes. Nature 300:611-615.

Palmiter, R. D., G. Norstedt, R.E. Gelinas, R.E. Hammer, and R.L. Brinster. 1983. Metallothionein-human GH fusion genes stimulate growth of mice. Science 222:809-814.

Palumbi, S. R. 1988. Molecular genetics of penaeid shrimp: characterization and utility of mitochondrial DNA variation within and between species. Abstract of paper presented at the 3rd international symposium on genetics in aquaculture, Trondheim, Norway, June 20-24 1988.

Pillai, M. C. and W.H. Clark, Jr. 1987. Oocyte activation in the marine shrimp *Sicyonia ingentis.* Journal of Experimental Zoology 244:325-329.

Pillai, M. C., F.J. Griffin, and W.H. Clark, Jr. 1988. Induced spawning of the decapod crustacean *Sicyonia ingentis*. Biological Bulletin 174(2):181-185.

Purdom, C. E. 1983. Genetic engineering by the manipulation of chromosomes. Aquaculture 33:287-300.

Ra'anan, Z. and D. Cohen, D. 1985. The ontogeny of social structure and population dynamics in the freshwater prawn *Macrobrachium rosenbergii* (de Man). Pages 277-311 *in* F. M. Schram and A. Wenner, editors. Crustacean issues. II. Crustacean growth. A. A. Balkema, Rotterdam.

Sagi, A., Z. Ra'anan, D. Cohen, and Y. Wax. 1986. Production of *Macrobrachium rosenbergii* in monosex populations: yield characteristics under intensive monoculture conditions in cages. Aquaculture 51:265-275.

Saiki, R. K., D.H. Gelfand, S. Stoffel, S.J. Scharf, R. Higuchi, G.T. Horn, K.B. Mullis, and H.A. Erlich. 1988a. Primer-directed enzymatic amplification of DNA with a thermostable DNA polymerase. Science 239:487-491.

Saiki, R. K., U.B. Gyllensten, and H.A. Erlich. 1988b. The polymerase chain reaction. Pages 141-152 *in* K. E. Davies, editor. Genome analysis, IRL Press, Washington.

Saiki, R. K., C.-A. Chang, C.H. Levenson, T.C. Warren, C.D. Boehm, H.H. Kazazian, and H.A. Erlich. 1988c. Diagnosis of sickle cell anemia and B-thallassemia with enzymatically amplified DNA and nonradioactive allele-specific oligonucleotide probes. New England Journal of Medicine 319:537-541.

Saiki, R. K., P.S. Walsh, C.H. Levenson, and H.A. Erlich. 1989. Genetic analysis of amplified DNA with immobilized sequence-specific olegonucleotide probes. Proceedings of the National Academy of Sciences USA 86:6230-6234.

Sbordoni, V., G. La Rosa, M. Mattoccia, M. Cobolli-Sbordoni, and E. De Matthaeis. 1987. Genetic changes in seven generations of hatchery stocks of the Kuruma prawn, *Penaeus japonicus* (Crustacea Decapoda). Pages 143-155 *in* K. Tiews, editor. Selection, hybridization and genetic engineering in aquaculture. Heenemann Verlag, Berlin, West Germany.

Schuler, J. 1988. Inserting genes affecting quantitative traits. Pages 198-199 *in* B. S. Weir, E. J. Eisen, M. M. Goodman, and G. Namkoong, editors. Proceedings of the 2nd international conference on quantitative genetics. Sinauer Associates, Sunderland, Massachusetts.

Smith, C. 1988. Potential for animal breeding, current and future. Pages 150-160 *in* B. S. Weir, E. J. Eisen, M. M. Goodman, and G. Namkoong, editors. Proceedings of the 2nd international conference on quantitative genetics. Sinauer Associates, Sunderland, Massachusetts.

Soller, M. and J.S. Beckmann. 1988. Genomic genetics and the utilization for breeding purposes of genetic variation between populations. Pages 161-188 *in* B. S. Weir, E. J. Eisen, M. M.Goodman, and G. Namkoong, editors. Proceedings of the 2nd international conference on quantitative genetics. Sinauer Associates, Sunderland, Massachusetts.

Stuber, C. W. 1989. Molecular markers in the manipulation of quantitative characters. *in* A. H. D. Brown, M. T. Clegg, A. L. Kahler, B. S. Weir, editors. Population genetics and germplasm resources in crop improvement. Sinauer Associates, Sunderland, Massachusetts. In press.

Telecky, T. M. 1982. The proportion of runt-fertilized females in two captive populations of *Macrobachium rosenbergii* and the behavior of runts associated with a courting bull and female. Masters thesis, University of Nevada, Reno. May 1982. pp. 1-53.

Telecky, T. M. 1984. Alternative male reproductive strategies in the Giant Malaysian Prawn, *Macrobachium rosenbergii*. Pacific Science: 372-373 (Abstract).

Yamazaki, F. 1983. Sex control and manipulation in fish. Aquaculture 33:329-354.

Zacarias, D. T. 1986. Determination of the relationship between larval and juvenile growth in *Macrobrachium rosenbergii* using genetically marked stock. MS Thesis, Department of Animal Sciences, University of Hawaii, Honolulu.

Zacarias, D., S.R. Malecha, L.E. Barck, E.R. MacMichael, S. Masuno, P. Ha-Tamaru, and D. Hedgecock. 1990. Biology and management of growth variation in the freshwater prawn, *Macrobrachium rosenbergii* I. Relationship between larval, juvenile and adult growth using allozymes as genetic tags. (manuscript in preparation).

PART III:

ECONOMICS CASE STUDIES

WILL US SHRIMP FARMS SURVIVE?: THE SOUTH CAROLINA EXPERIENCE

Raymond J. Rhodes

ABSTRACT

In 1987 a record level of shrimp, 583 million pounds, was imported into the U.S.A. while the production of foreign cultured shrimp continues to expand. Given this situation and other factors (e.g., rising feed costs), there is concern that the development of marine shrimp farming in the U.S.A. will be stifled. The current situation and outlook of shrimp farming in the continental U.S.A. is examined using South Carolina as an example.

Based upon earthen pond culturing techniques, the financial performance of a 32-acre (13-ha) intensive marine shrimp farm was simulated for South Carolina. Over a ten year period, annual income and cash flow statements were projected. The 10-year net present value, internal rate of return, and the "average" breakeven price were calculated for various scenarios. Based upon the financial analysis and other information, it appears that South Carolina intensive marine shrimp farms can survive and generate positive returns to their owners and/or investors.

The current major market segments for selling South Carolina cultured shrimp remain exposed to the commodity-oriented pricing of most U.S. shrimp. The marketing mix of South Carolina shrimp farms must include "niche" market segments which will insulate a significant amount of domestic cultured shrimp production from the fluctuation of regional and national supply induced price changes.

INTRODUCTION

As an observer and analyst of aquaculture developments in the U.S. and other countries, it is a pleasure to have the opportunity to share my views on the development of marine shrimp farming in the U.S. Discussing shrimp farming is especially interesting to an economist because both the physical and socio-economic environment in each country have obviously shaped the development of shrimp farming. In a sense, the biology of various shrimp species and basic culturing techniques are the sculptor's "wood" and how this "wood" is "sculpted" into shrimp farming in a given country is influenced not only by the shrimp farmer but also by the environment in which he must sculpt. This is no less true in a developed country like the U.S.

In this discussion, I will not dwell on the current forces shaping the various shrimp market segments in the U.S. because many others have documented and analyzed the current situation and future outlook for the U.S. (e.g., N.M.F.S. 1988; Filose 1988). You do not need to be a market analyst to realize that the rapid expansion of U.S. shrimp supplies, mainly due to imports, has impacted shrimp consumption (Fig. 1) and prices. Consequently, this paper concentrates on providing general observations on factors influencing the current (1988) financial performance of South Carolina marine shrimp farms, plus my views on how these and other factors might influence the financial future of commercial shrimp farming in South Carolina. I will give my answers to the following questions: Are South Carolina shrimp farms currently profitable; and (2) Can South

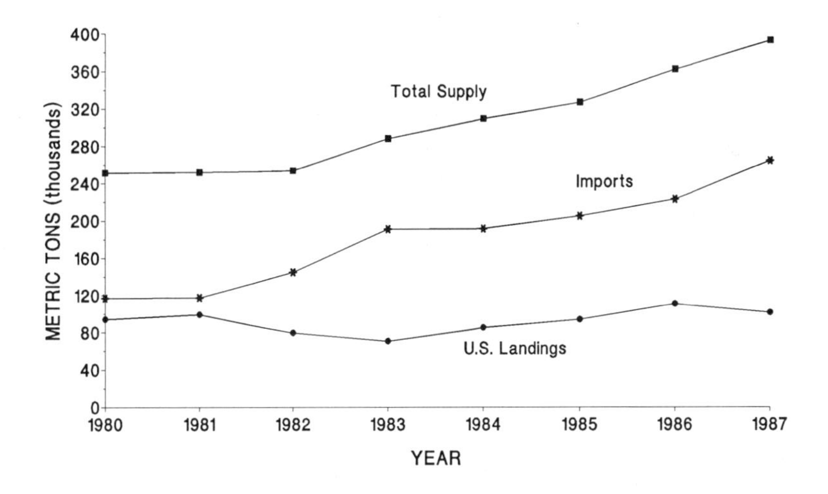

Figure 1. U.S. shrimp imports and domestic landings, 1980-1987.

Carolina farms continue as ongoing enterprises and generate a "reasonable" profit for their owners and/or investors? My opinions on the current and future economic financial performance in South Carolina need to be viewed in the context of how the public and private sectors in one U.S. state are trying to reshape and sculpt the "wood" of basic shrimp farming techniques into a profitable and sustainable enterprise in the context of South Carolina's physical and socio-economic environment. Moreover, the generalization developed relative to South Carolina should be only cautiously extrapolated to other U.S. regions, let alone to other countries. Last but not least, I will not address environmental regulations currently constraining shrimp farming development in the U.S. This important topic has been addressed by others.

CURRENT STATUS OF SOUTH CAROLINA FARMS

Hopkins (1988) reported that South Carolina accounted for about 17% of the 2.3 million pounds of shrimp farmed in the U.S. during 1987. Hopkins (1988) also noted that some U.S. shrimp farms were making modest profits in 1987 and others were "...just hanging on." Based upon my observations in South Carolina for the 1988 season, Hopkins' generalization still applies to South Carolina, but I provide some specific details to "just hanging on." South Carolina shrimp farms can be categorized into two general groups based upon the motivation of the owners: full-scale commercial enterprises and private pilot-scale farms. The farms working at or near commercial scale during the last three seasons may have generated positive cash flows but are not generating a positive return on investment relative to the initial investment. The financial extrapolation of the pilot-scale experience, excluding the experimental work at the Waddell Mariculture Center (WMC), suggests that some farming techniques could be profitable at proposed full-scale level. These projections are discussed elsewhere in this paper. Since shrimp farming in the U.S., and especially South Carolina, is a new industry, it should be no surprise that farms are having difficulties. Remember in the U.S. during 1985, over 57,000 companies closed their doors. New small business enterprises always have difficulties, regardless of the technology and market situation.

FACTORS AFFECTING PROFITABILITY

The most germane question is what factors are constraining the profitability of South Carolina shrimp farming? As alluded previously, S.C. shrimp farming development has been subject to the "learning curve" problems of adapting shrimp farming techniques to South Carolina's environment. The WMC has played a critical role in modifying proven techniques (Sandifer et al. 1987) from other countries like Taiwan and communicating these techniques to the private sector. Based upon my three years of observations, it appears that the WMC culturing techniques can be effectively transferred to South Carolina's private sector. The adaptation and scaling-up of these techniques to specific sites then ultimately becomes subject to the intrinsic judgement and innovation of aquaculture entrepreneurs and others in the private sector.

Besides the normal technological and management learning curve problems endemic to any new production technology, U.S. marine shrimp market prices declined faster than anticipated since 1986 (Fig. 2). Local wholesale and ex-vessel shrimp prices also declined due to national and

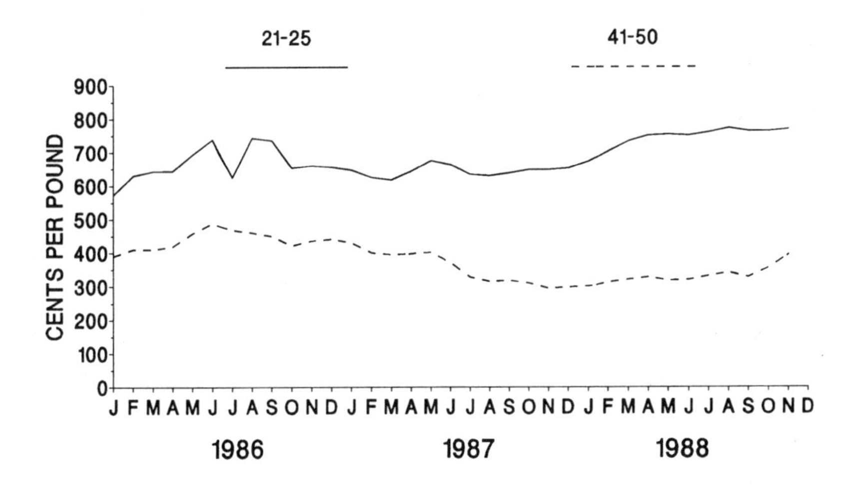

Figure 2. Ecuadorean (heads-off) white shrimp wholesale prices at New York, 1986, Jan.-Nov. 1988.

regional price effects. These major shrimp price fluctuations are to be expected (Rhodes et al. 1987) but the 1987 downturn amplified other problems faced by shrimp farms. Wholesale shrimp prices did improve some in 1988 (Fig. 2), but the 1989 price situation is unclear given the expanding cultured shrimp production in Asia.

Feed costs have increased significantly since 1986. Between 1986 and 1988, the price of shrimp grow-out feed increased nearly 25%, reaching a level of about $0.31/lb ($0.68/kg). Feed expenses can easily constitute 30% of shrimp farming expenses (Rhodes et al. in preparation). The increase in U.S. shrimp feed cost was partially attributable to the increase in fish meal prices (Fig. 3). Fish meal prices increased nearly 30% between May 1987 and May 1988 (Fig. 3). In addition to rising feed prices, there was also concern that the price of hatchery produced post-larvae would increase significantly in 1989.

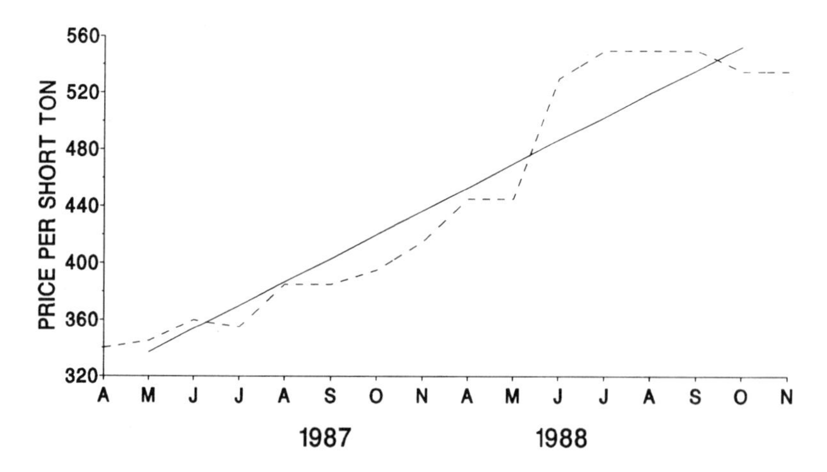

Figure 3. U.S. fish meal market prices, 1987-1988 (April to November).

Ironically, South Carolina shrimp farms, which are still developing their production techniques and markets, are caught in a cost-price "squeeze" similar to the type the South Carolina shrimp trawler fleet faced in 1974 when fuel prices escalated and shrimp prices declined. For shrimp farmers, the feed represents a critical input cost not directly related to the cost of energy. Last but not least, expansion plans based on venture capital were probably hindered by the October 1987 downturn of the New York stock market. This situation not only hampered the direct investment in U.S. shrimp farming, but may have seriously retarded the development of the infrastructure (e.g., aquaculture feed mills, shrimp hatcheries, etc.) necessary to the sustained growth of Southeastern shrimp farming.

OUTLOOK FOR SHRIMP FARMING

The second question was: Can S.C. farms continue as ongoing enterprises and generate a "reasonable" profit for their owners and/or investors? Based upon current information, it appears that S.C. shrimp farms can generate profits in future years, but this will depend upon a set of conditions, some of which are not under the control of the aquaculture entrepreneur. The major uncontrollable conditions[1] include no major decline in wholesale shrimp prices, (e.g., 31-35 through 56-60 heads-off count size shrimp) for at least the next two or three years and a stabilization of major input prices, especially "seed" (postlarvae) and feed. South Carolina farmers are already concerned with the availability and cost of shrimp postlarvae for the 1989 season. In addition, shrimp farmers are encouraging the entry of more shrimp feed suppliers, but none are expected to enter the market in 1989. Given these conditions, it is not surprising that several farms are planning to culture other species (e.g., clams and oysters) in addition to shrimp.

IMPORTANCE OF INTENSIFICATION

Private pilot-scale shrimp farming efforts in South Carolina indicate that yields over 10,000 lbs/acre (11,210 kg/ha) can be achieved. In addition, research at the WMC by Sandifer et al. (1988) suggests that there is no apparent major density-dependent effect on shrimp harvest size, with increasing density from 20 to at least as high as 100/m^2. They predicted that harvests of over 9,000 lbs/acre/crop (about 10,000 kg/ha/crop) could be routinely harvested on U.S. farms within the next ten years.

---

[1] This assumes that current regulatory constraints will not be a major hinderance to the expansion of small intensive farms (e.g., less than 100 acres).

A preliminary[2] financial analysis of single-crop shrimp farming in South Carolina (Rhodes et al. in preparation) suggests that intensification (i.e., PL stocking densities of over 40/m^2) may also be the most profitable approach. Based upon a direct PL stocking density of 100/m^2 and an average harvest size of 15 g shrimp (Table 1) and other assumptions (Tables 1 and 2), it appears that a positive net cash flow (Table 3) can be generated over a ten-year period by a 32-acre (13 ha) intensive shrimp farm, resulting in a projected after tax internal rate of return (IRR) of 18% (Table 4). Based upon these preliminary assumptions, the breakeven[3] size and price were about 12.5 g (Fig.4) and $ 1.93/lb ($ 4.25/kg) (Fig.5), respectively. These financial projections for intensive shrimp farming also demonstrate that profitability is most sensitive to changes in heads-on shrimp prices (Fig. 6) when compared to harvest size and PL prices (Fig. 7)

Table 1. Major production assumtions for the preliminary financial analysis of an intensive marine shrimp farm in South Carolina, 1988.

	32-acre (13-ha) Marine Shrimp Farm in South Carolina
Pond size:	2 Acres (0.80 ha)
Number of Ponds:	16
Stock Density:	404,700/a (100/m^2)
Aggregate Survival Rate:	75%
Average Harvest Size:	16g
Average Yield:	10,040/a (11,250 kg/ha)
Feed Conversion Ratio:	2.4: 1
Paddlewheel Aeration:	10 hp/a (25 hp/ha)

DEVELOPMENT OF NICHE MARKETS

When discussing the evolution and development of aquaculture in Europe and the Americas, Sandifer (1988) concluded that the key to expanding aquaculture production is marketing. The same generalization also applies to the future development of shrimp farming, especially given the

[2] This paper was only intended tp provide a brief overview of shrimp farming production economics given the commercial trend toward intensification.

[3] The lowest projected income year (Year 2) was used when estimating the breakeven price and harvest size.

Table 2. Major financial assumptions for the preliminary financial analysis for an intensive marine shrimp farm in South Carolina, 1988.

	32-acre (13-ha) Marine Shrimp Farm in South Carolina
Land Purchased:	$13,500/a ($8,645/ha)
PL ("seed") cost	$10/1,000
Feed Cost:	$0.31/lb ($0.68/kg)
Electricity:	$0.07/KWH
Market Price (Head-on):	$2.15/lb ($4.74/kg)
Salary & Wage Labor:	$86,400/yr
Total "Start-up" Investment*:	$458,000
Total Capital & Operating Investment**:	$1.3 million

* Excluding land and operating capital

** Including land and operating capital.

Table 3. Preliminary projected income statement for a hypothetical intensive farm in South Carolina.

	32-acre (13-ha) Marine Shrimp Farm (in Thousands)
Projected Gross Sales:	$691,000
Direct Operation Expenses:	$443,000*
Projected Gross Profit:	$248,000
Other (Fixed) Operating Expenses:	$174,000**
Total Projected Expenses:	$617,000
Projected Net Income:	$74,000
Projected Total Cost/lb:	$1.92/lb ($4.23/kg)

* Feed costs were 54% of the category.

** Salaries & wages were 50% of this category.

Table 4. Preliminary projected 10-year internal rate of return (after taxes) for a hypothetical 32-acre intensive shrimp farm in South Carolina with a projected average yield of 11,250 kg/ha.

Heads-on Market Price $/lb	10-Year IRR Purchased Land	10-Year IRR Leased Land
1.85	-10%	-7%
* 2.15	18%	29%
2.25	24%	37%

*Base case price.

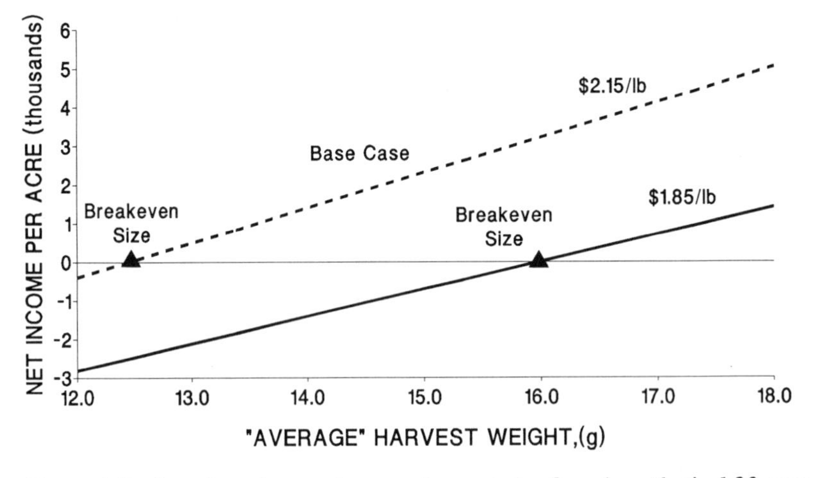

Figure 4. Projected net income/acre vs. harvest size for a hypothetical 32-acre intensive shrimp farm in South Carolina.

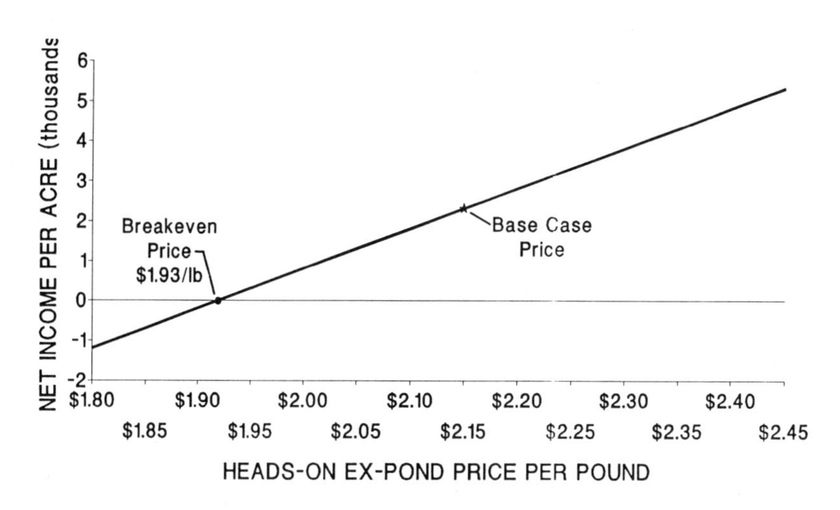

Figure 5. Projected net income/acre vs. market price for a hypothetical 32-acre intensive shrimp farm in South Carolina.

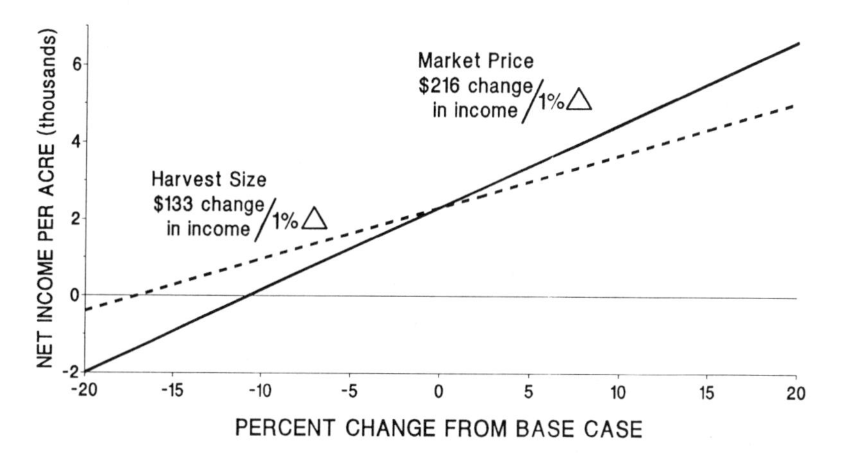

Figure 6. Projected net income per acre vs. a percent change in market price or harvest size for a hypothetical 32-acre intensive shrimp farm in South Carolina.

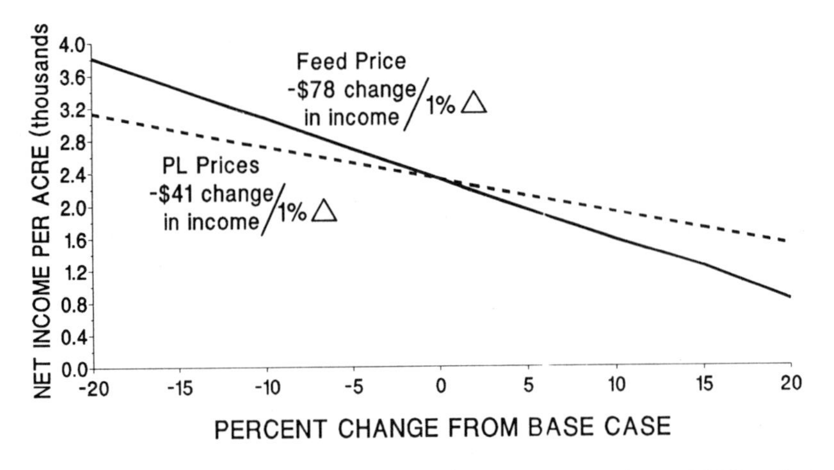

Figure 7. Projected net income per acre vs. a percent change in seed (PL) or feed prices for a hypothetical 32-acre intensive shrimp farm in South Carolina.

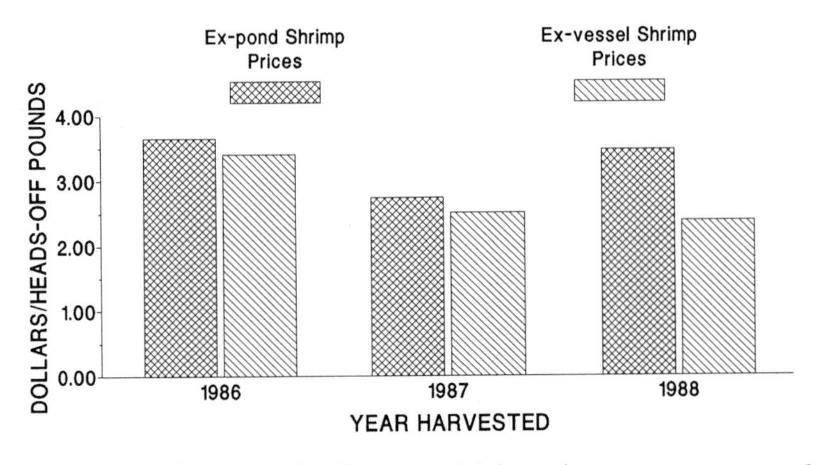

Figure 8. Estimated South Carolina ex-pond shrimp prices vs. average ex-vessel prices (October) (41-45 heads-off shrimp/lb), 1986-1988.

relative high sensitivity of farming income to changes in shrimp prices (Fig. 6). Some progress has been made in securing higher ex-pond heads-on prices for cultured shrimp based upon a preliminary comparison to South Carolina ex-vessel shrimp prices (Fig. 8). Whether this price difference is really a "price premium" or just a price differential associated with selling directly to primary and/or secondary wholesalers is a moot point if the major market segments for selling South Carolina cultured shrimp remain exposed to the commodity-oriented pricing of most U.S. shrimp.

The marketing mix of shrimp farms must include "niche" market segments which will insulate a significant amount of South Carolina cultured shrimp production from the fluctuation of regional and national supply-induced price changes. For example, a potential niche segment includes the "skim" harvesting and distribution of live shrimp for coastal fishermen in the Carolinas and Georgia, (Table 5). In 1982, the total amount of live bait shrimp in South Carolina alone was estimated at about 400,000 pounds (Moore 1982). This is a small market segment that is isolated from foreign shrimp producers and is only seasonally serviced by small specialized fishing operations. We are planning to evaluate these and other potential markets for South Carolina shrimp producers with funding support from the Gulf Coast Research Laboratory Consortium and the Division of Marine Resources.

Table 5. Some potential "niche" markets for South Carolina marine shrimp farms.

- Direct pond-side sales to consumers and small wholesalers
- Direct harvesting by consumers (Fee Shrimping)
- Live bait distribution
- Heads-on shrimp sales in major shopping areas.

CONCLUDING REMARKS

Some may interpret my analysis of the S.C. private shrimp farming situation and outlook as a prologue to a pending decline. My only intent is to provide a quick "snapshot" of the South Carolina shrimp farming scene, not a final portrait. Moreover, the current picture suggests that the private sector is evolving through the usual problems faced by any new agricultural production technology including perhaps the neglect of marketing tasks. The increase in imported cultured shrimp does constitute an obvious challenge to the continued expansion and survival of U.S. cultured shrimp producers. Ironically, South Carolina's small (9) producers may actually have more potentially profitable market segments to mitigate the expected price effects of expanding shrimp production compared to some shrimp farming operations in other countries.

LITERATURE CITED

Filose, J. 1988. The North American market perspective. Pages 20-27 *in* Shrimp '88. INFOFISH, Kuala Lumpur, Malaysia.

Hopkins, J.S. 1988. Shrimp farming in the United States. Presented at Shellfish Institute of North America Annual Convention, Feb. 28-March 2, 1988 Charleston, South Carolina, U.S.A. (unpublished manuscript).

Moore, C. J. 1982. A description of natural marine bait utilization in South Carolina. Pages 37-42 *in* Proceedings of the marine natural bait industry workshop. S.C. Sea Grant Consortium, Proc. No. 1 (SCSP-PR-82-01), Charleston, South Carolina, U.S.A.

N.M.F.S. 1988. Aquaculture and capture fisheries: Impacts in U.S. seafood markets. U.S. Department of Commerce, Washington, D.C., U.S.A.

Rhodes, R.J., P.A. Sandifer, and J.M. Whetstone. 1987. A preliminary financial analysis of semi-intensive penaeid shrimp farming in South Carolina. South Carolina Marine Resources Center, Technical Report 64, Charleston, SC, U.S.A., 23 pp.

Rhodes, R.J., P.A. Sandifer, J.S. Hopkins and A.D. Stokes. In Preparation. A preliminary financial analysis of intensive marine shrimp farming in South Carolina. South Carolina Marine Resources Center, Technical Report. Charleston, SC, U.S.A..

Sandifer, P.A. 1988. Aquaculture in the West: a perspective. Journal of the World Aquaculture Society 19(2): 73-84.

Sandifer, P.A., J.S. Hopkins and A.D. Stokes. 1987. Intensive culture potential of *Penaeus vannamei*. Journal of the World Aquaculture Society 18:94-100.

Sandifer, P.A., J.S. Hopkins and A.D. Stokes. 1988. Intensification of shrimp culture in earthen ponds in South Carolina: progress and prospects. Journal of the World Aquaculture Society 19(4): 218-226.

THE POTENTIAL IMPACT OF SHRIMP MARICULTURE DEVELOPMENT ON THE ECONOMY OF SOUTH CAROLINA: A MODEL BASED ON AVAILABLE LAND RESOURCES

Bruce J. French, Linos Cotsapas and Miles O. Hayes

ABSTRACT

The coastal land resources of the southeastern United States are under intense pressure for future development by a number of competing interests. Some of these activities threaten the water quality of some of the most pristine estuarine systems in the nation, particularly those in South Carolina and Georgia. Shrimp mariculture provides a viable alternative to these types of land uses, because, while it provides a high rate of return on economic investments, its impact on the estuarine ecosystem is minimal.

A preliminary analysis of the coastal lands of South Carolina was conducted to determine the potential magnitude of a local shrimp mariculture industry. Approximately 56,000 ha located on the Pleistocene highland adjacent to brackish water were identified as being suitable for commercial shrimp farming. The economic impact of developing 5, 10, 25, and 50 % of this land resource was forecasted. Assuming that hatcheries, feed mills, processing plants, and support services would be contained within the state, the benefit to the state's economy could reach $2.4 billion annually. In addition, a maximum of 13,900 jobs would be created. An activity with a potential economic impact of this magnitude will possibly become a major competitor for the valuable coastal land resources of South Carolina and its neighboring coastal states.

INTRODUCTION

In the summer of 1987, RPI International, Inc. undertook an in-house effort to forecast an ideal infrastructure for a fully developed marine shrimp culture industry in South Carolina. This study in resource allocation and industry development was presented from a commercial, entrepreneurial perspective. It was reviewed by 12 government employees knowledgeable about aquaculture and was presented to the Governor's Office in March 1988. Hopefully, the process employed in this study will serve as a template for preparing similar industry development plans for other species and locales.

METHODS

The intent of the study was to forecast the structure of a shrimp culture industry in South Carolina, using the Mississippi farm-raised catfish industry as a model. A marine shrimp culture industry basically consists of the following major business components:

- Shrimp hatcheries.
- Shrimp farms.
- Processing plants.
- Feed mills.
- Grain farms.
- Supporting sectors (e.g., transportation, utilities, equipment suppliers, aquaculture consultants, etc.)
- Government services.

A database was collected related to the various components of a shrimp culture industry in South Carolina. It included information on the following topics:

- Historical records on the development of the farm-raised catfish industry in Mississippi.
- Economic model of the farm-raised catfish industry.
- USA and world shrimp markets.
- Nutritional requirements and feed formulations for penaeid shrimp.
- Feed milling operations in South Carolina.

- Financial analysis on the production of penaeid shrimp in South Carolina.
- Land use in the South Carolina coastal zone.
- Soil survey of eight coastal counties in South Carolina.
- Surface water quality of the South Carolina coastal zone.

From this database, a set of assumptions was prepared for constructing the industry components and forecasting the impact on the South Carolina economy.

DESCRIPTION OF COASTAL SOUTH CAROLINA

South Carolina encompasses a land area of approximately 7.82 million ha. It lies between 32 and 34 ° North latitude and has a temperate climate. Geographic features of the state include upland mountains, a central piedmont region, and a "low country" coastal plain (Fig. 1).

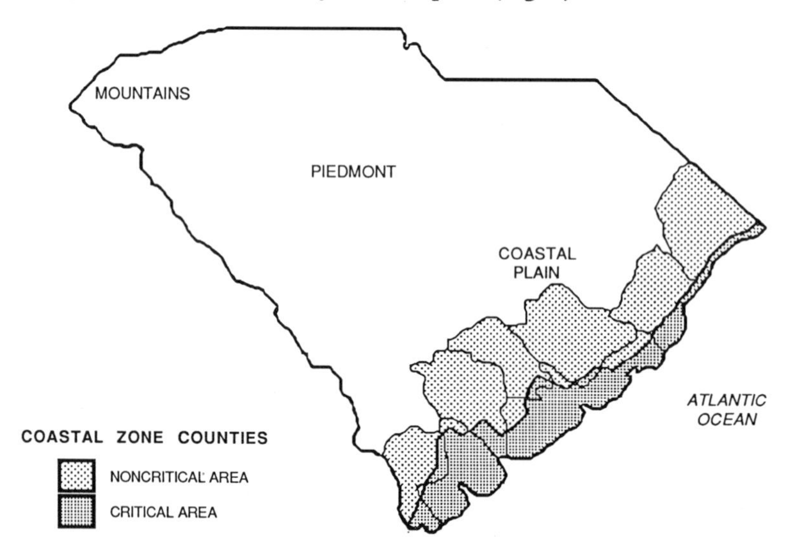

Figure 1. Land zonation in South Carolina, highlighting the eight counties in the coastal zone and the critical area within these counties.

The convoluted coastline on the Atlantic Ocean is about 350 km in length, if one includes the shorelines of the eight major estuarine systems in the state. Some of these estuaries are significantly influenced by freshwater rivers draining the interior, whereas others are marine systems with little freshwater influence. Extensive salt marshes exist along the coast owing partially to the large tidal range (± 2 m).

The coastal zone is divided into 8 counties with a total area of 1.83 million ha. It is divided into critical (506,265 ha) and and noncritical (1,320,192 ha) areas. The critical areas are tidally influenced, and the noncritical areas are uplands influenced mainly by freshwater hydrology.

In 1987, 818,900 people or 24 % of the State's 3.4 million population lived in the coastal zone. Land use within the coastal zone was analyzed by state government in 1979. Eleven land-use types were identified. For this study, use was divided into two categories:

Group A
- Urban or built-up land
- Impoundment
- Forested wetland
- Salt marsh
- Brackish marsh
- Freshwater marsh
- Water

Group B
- Agriculture
- Range lands
- Forest lands
- Barren lands

Because shrimp culture requires brackish water, the majority of the state's industry must be developed in the critical area of the coastal zone. Only Group B land-use types are suitable for the development of shrimp farms. Six of the coastal counties have lands which are physically suitable for shrimp farming with a total area of 1.381 million ha. The critical area within these 6 counties is 488,256 ha (35.35 %), and the Group B type within the critical area is 195,111 ha (14.13 %). Obviously, shrimp farm development will compete directly with other land-use activity for the limited property within the critical zone.

The S.C. Coastal Council is responsible for regulating human activities within the coastal zone. Between 1980 and 1985, an annual average of 350 permits have been issued within the critical area. Most of these are for private residential boat docks and bulkheads. If we assumed that each residential property averages only 60 m wide, then at least 21 km of shoreline would be removed from the land resource base of the critical area each year.

Obviously, the total land area of a fully matured state shrimp culture industry will be heavily dependent upon the rate at which farm development captures the available, limited land resource.

LAND SELECTION

After a review of the data, a basic map survey of the South Carolina coastal zone was conducted to identify land that was potentially suitable for shrimp culture. Because intensive culture methods have been developed employing plastic-lined, concrete-walled, and earthen-clay ponds, no limitations were set for soil type.

Land areas less than 30 ha were not included in the resource base because of the requirement for an economy of scale. Also, property selected had to have access to surface waters in excess of 10 ppt salinity and have an elevation of 0-5 m above high tide.

Of the 8 coastal counties, 2 were eliminated from the land selection process. Horry County has a narrow critical area and is highly developed as a tourist resort. Dorchester County has minimal suitable land existing in the critical area.

From our study, the 6 counties were found to have approximately 56,232 ha of Group B land physically suited for shrimp farm development

Table 1. Counties found to contain Group B-type land physically suited for shrimp farm development.

County	Group B-Type Land (in ha)	Suitable Land for Shrimp Farms (in ha)
Jasper	17,740	3,715
Beaufort	55,929	12,473
Colleton	13,667	4,701
Charleston	86,346	22,881
Berkeley	8,311	5,935
Georgetown	13,118	6,528
Totals	195,111	56,232

(Table 1). This represents about 4.1 % of the total area of these counties, 11.5 % of the total critical area, and 28.8 % of the available Group B land resource.

Because the geographic data included information sources from various years, it was unrealistic to assume that 100 % of the identified lands remained available for development. Although physically suitable, some of the selected properties obviously were contained in government lands or possibly existed in other permanent land use.

In the absence of detailed, current land-use data, it was decided to project the economic impact of the industry at 4 levels of land resource development: 5, 10, 25, and 50 % of the total 56,232 ha of land found to be theoretically suitable for shrimp farming.

If 50 % (28,116 ha) of the shrimp farm land resource identified were developed, this land use would represent about 2.04 % of the total area in the 6 counties, 5.8 % of the total critical area, and 14.4 % of the total Group B area.

SHRIMP FARM YIELD

At the time of this study, only 1986 production results were available from research facilities in the United States. Although research yields had climbed to 7,000 kg/ha/crop, we decided to use a conservative yield attainable by the local commercial industry. The live-weight yield used in the forecast was 3,362 kg/ha, produced in one crop annually and at the four levels of land resource development (Table 2).

Table 2. Shrimp live-weight yield at four levels of land resource development.

Percent Land Development	Number of Hectares	Live-Weight Yield (in million kg/yr)
5%	2,812	9.453
10	5,623	18.905
25	14,058	47.263
50	28,116	94.526

SHRIMP FARM ECONOMICS

In forecasting the associated development costs of the shrimp farming sector of the industry, the set of assumptions presented by Rhodes et al. (1987a) was modified to reflect the live-weight yield used in this study (Tables 3 and 4).

For employment rates, it was estimated that for each 80.9 ha of shrimp farm, 4 full-time and 9 seasonal jobs would be required for operations.

FEED MILLS

Several shrimp feed formulations were examined (New 1976; Fenucci et al. 1980; National Academy of Sciences 1983), and a hypothetical commercial feed formula used (Table 5) in calculating grain demand and in projecting certain industry parameters. This formula is hypothetical, and it is not intended to be used as a commercial shrimp feed formula.

Table 3. Estimated capital investment per hectare and at four levels of land resource development.

Pond system construction	$ 7,015
Building and equipment	7,079
Operating capital and reserve fund	13,870
Total Investment/ha	$ 27,964

Percent Land Development	Number of Hectares	Capital Investment (in $ million)
5%	2,812	$ 78.6
10	5,623	157.3
25	14,058	393.1
50	28,116	786.3

Table 4. Annual shrimp farm income statement per hectare and total shrimp farm area income at four levels of land resource development.

Farm sales per kilogram of tails	$ 7.94
Annual farm sales	16,812
Annual operating costs	
Direct costs	8,641
Operating costs	5,231
Total annual operating costs	13,872
Pretax profit	2,940
Estimated taxes	1,176
Net profit after taxes	$ 1,764

Percent Land Development	Annual Sales (in $ million)	Operating Costs (in $ million)	Net Profit (in $ million)
5%	$ 47.3	$ 39.0	$ 5.0
10	94.5	78.0	9.9
25	236.3	195.0	24.8
50	472.7	390.0	49.6

Table 5. Hypothetical shrimp feed formula.

Ingredient	%	Ingredient	%
Corn	20	Fish oils	5
Wheat	15	Squid meal	5
Soy beans	5	Whey	2
Rice bran	10	Lecithin	2
Shrimp meal	20	Vit./Min. mix	3
Fish meal	11	Alginate	2

Table 6. Annual feed sales at four levels of land resource development.

% Land Development	Annual Feed (in tons)	Sales (in $ million)
5	18,904	$ 9.6
10	37,808	19.2
25	94,521	47.9
50	189,042	95.9

Table 7. Daily output from feed mills at four levels of land resource development.

% Land Development	Feed Mills	Daily Output (in tons)
5	1	116
10	1	232
25	3	579
50	4	1,158

Table 8. Employment ratios per ton of milled feed.

Office jobs	0.017
Mill management	0.021
Feed production	0.083
Transportation	0.017
Total Jobs/Ton	0.138

Annual feed and ingredient demand were determined by assuming that shrimp would convert feed at an FCR of 2:1. The assumption was made that local feed mills would produce the pelleted formula for $0.507/kg. Annual feed sales were then forecasted (Table 6).

To meet the demands of shrimp farms, it was assumed that mills would produce shrimp feeds for 180 days each year. Each mill would have the capacity to produce up to 290 tons per day (Table 7).

Employment requirements for the mills were determined through a telephone survey of feed mills in the state. From this survey, it was possible to establish a job ratio per ton of feed milled daily using the Goldkist Mill in Gaston, SC, as a model (Table 8).

GRAIN PRODUCTION

Ideally, three ingredients of the feed formula—corn, wheat, and soybeans—could be produced entirely from SC agriculture. The annual demand of these three ingredients is reflected in Table 9. The state's grain belt, which includes 18 counties, lies adjacent to the coastal zone.

Excluding the deficiency payment received by farmers in the state, the market price and yield/ha for corn, wheat, and soybeans were averaged from 1970-1985 records (Table 10) (S.C. Agricultural Statistics Service 1987a,b; Cooperative Extension Service 1988; Pugh 1988).

The average grain farm in this area is 81 ha. Total farm area and potential sales are reported in Table 11. Each farm was assumed to employ 1 full-time farmer and 2 seasonal laborers.

SHRIMP PROCESSING PLANTS

By comparing the processing status of the USA Gulf shrimp industry and the Mississippi catfish processing structure, the processing requirement for the South Carolina shrimp farming sector was constructed. Under our scenario, all farm-raised shrimp would be processed at plants located in the state. Each plant would have an average processing capacity of 4,600 tons

Table 9. Annual grain demand for corn, wheat, and soybeans to be used in the formulation of a shrimp diet at four levels of land resource development.

% Land Development	Annual Grain Demand (in tons)		
	Corn	Wheat	Soybeans
5	3,781	2,836	945
10	7,562	5,671	1,890
25	18,904	14,178	4,726
50	37,808	28,356	9,452

Table 10. Average market price and yield per hectare of corn, wheat, and soybeans for 1970-1985 in South Carolina.

Ingredient	Kilograms/ Hectare	Price/ Kilogram
Corn	2,887	$ 0.07
Wheat	1,681	0.09
Soybeans	1,143	0.18

Table 11. Potential farm area, number of farms, and total sales from corn, wheat, and soybean farms in South Carolina at four levels of land resource development.

% Land Development	Total Hectares	Total Farms	Total Sales (in $ million)
5	3,815	47	$ 0.675
10	7,630	94	1.351
25	19,074	236	3.377
50	38,149	472	6.753

Table 12. Number of processing plants, tail weight produced, and sales generated at four levels of land resource development.

% Land Development	Number of Plants	Tail Weight (in tons)	Sales (in $ million)
5	1	5,969	$ 57.78
10	3	11,937	115.55
25	6	29,843	288.88
50	13	59,685	577.75

of shrimp tails per season. Processing would occur during a 90-day fall season. Each plant would process an average of 50.9 tons per day (Table 12), servicing producers within a 60-km radius. Plants would sell an average tail size of 31/35 count/pound for $9.68/kg.

Head waste generated by processing was estimated to yield 6 % dry weight of the whole live-weight shrimp. This by-product would provide between 567 and 5,673 tons of dry shrimp meal or about 15 % of the annual demand for the ingredient in shrimp feed formulation. At $275 per ton, annual meal sales by processors would range from $156,000 to $1,560,000 depending on the level of land resource development.

Processing plant operations were estimated to require 8 full-time employees and an additional 200 seasonal workers per facility.

INDUSTRY EMPLOYMENT

Statistical information on employment by the farm-raised catfish industry in Mississippi has been reported by two sources. One states that 1 job was created for each 1.74 ha of catfish farm ponds in production. The other uses 3 economic models which show that each catfish farm job created resulted in an employment multiplier of 2.09, 2.61, or 6.97 jobs state-wide.

For this study, it was assumed that 1 job would be created in all sectors of the industry statewide for each 2.02 ha of shrimp farm constructed (Table 13). The numbers of jobs created on shrimp farms, grain farms, feed mills, and processing plants are reported in Table 13.

Table 13. Jobs created in the shrimp farm, grain farm, feed mill, processing plant, and all other sectors at four levels of land resource development.

% Land Development	Number of Hectares	Total Jobs (Full-Time and Seasonal)
5	2,812	1,390
10	5,623	2,779
25	14,058	6,948
50	28,116	13,895

% Land Development	Shrimp Farms	Grain Farms	Feed Mills	Processing Plants	Other Sectors
5	456	141	16	208	568
10	911	283	32	624	929
25	2,278	707	80	1,248	2,634
50	4,557	1,415	159	2,704	5,060

Table 14. Effect of a 5X economic multiplier on shrimp farm sales in South Carolina at four levels of land resource development.

% Land Development	Farm Sales (in $ million)	5X Economic* Multiplier (in $ million)
5	$ 47.3	$ 236.4
10	94.5	472.7
25	236.3	1,181.8
50	472.7	2,363.5

* Editor's note: Most economists apparently accept multipliers ≤ 2 for this application.

Based on the above assumptions, the shrimp farms, grain farms, feed mills, and processing plants account for 63 % of all jobs created statewide by the industry. All other sectors represent 37 % of all jobs created.

ECONOMIC BENEFITS

Although we feel fairly comfortable with the set of assumptions used up to this point, when it came to determining an economic multiplier for the shrimp farm sector, it was difficult to ascertain which of the numerous models used by economists was valid for this case. In canvassing several economists and studies, it was suggested that industries of this type could have economic multipliers ranging from less than 1.5 to 7.7.

One economist, who preferred to remain anonymous, felt that if all the inputs of an industry could be produced in-state with a minimum of leakage, then an ideal, maximum multiplier that such an industry could expect would be 5 times farm sales. Considering that the South Carolina shrimp culture industry was intended to be a home-grown, grass-roots industry, it was desirous to forecast the maximum economic benefit to the state. The effect of this multiplier on shrimp farm sales is presented in Table 14. Nevertheless, most aquacultural economists appear to accept a multiplier not greater than two.

We also prepared a matrix on farm sales, farm payroll income, and farm jobs versus 4 multiplier values (2.0, 3.0, 4.0, and 5.0). This matrix was performed at the four levels of land resource development. The economic impact of farm sales ranged from $94.5 million to $2.4 billion. Payroll income impact ranged from $10.9 million to $272.9 million. Based on the farm jobs, a range of 911 to 22,772 jobs are created industry-wide.

Assuming that (1) the multiplier of 5 remains valid over time, (2) the land resource can be developed in annual increments of 2,812 ha (5 % of the resource), and (3) a maximum of 28,116 ha (50 % of the land resource) is developed over a 10-year period, the cumulative economic benefits would be substantial. These benefits are presented in Table 15.

Table 15. Economic benefits to South Carolina.

Year	Percent Resource Developed	Economic (5X)* Benefit (in $ billion)
1	5%	$ 0.24
2	10	0.47
3	15	0.71
4	20	0.95
5	25	1.18
6	30	1.42
7	35	1.65
8	40	1.89
9	45	2.13
10	50	2.36
10-Year Cumulative		$ 13.00
20-Year Cumulative		$ 36.63

* Editor's note: Most economists apparently accept multipliers ≤ 2 for this kind of application.

Table 16. Estimated tax revenues generated by the industry at four levels of land resource development.

% Land Development	Pretax Profits (in $ million)	Federal Taxes (in $ million)	State Taxes (in $ million)
5	$ 47.3	$ 16.1	$ 3.4
10	94.5	32.1	6.7
25	236.4	80.4	16.8
50	472.8	160.8	33.6

TAXES

To date, there are about 120 ha of commercial marine shrimp farms in South Carolina which are benefiting from the state's research and extension efforts. The state's mariculture research center was constructed in 1984 at a cost of approximately $4 million. Facilities such as these require taxes to build and operate.

By examining the economic benefit of shrimp farm sales (5X multiplier) and making the assumption that all sectors will enjoy a 20 % pretax profit, the state and federal tax revenues can be estimated.

Federal tax revenues were estimated on 34 % of pretax profits. State taxes were estimated on the sum of a 3 % payroll income tax and a 6 % corporate profit tax. Tax revenues were forecasted at four levels of land resource development (Table 16).

Assuming that 5 % of the land resource can be developed into shrimp farms annually and a maximum of 50 % of the total is developed in 10 years, then the cumulative federal tax revenue amounts to $4.42 billion and the South Carolina state tax revenue is $2.72 billion. The 20-year totals are $12.45 billion and $7.66 billion, respectively.

We believe that legislators would be encouraged to finance the development of coastal aquaculture industries more strongly if they had a better understanding of the potential economic benefits.

SUMMARY AND DISCUSSION

By considering the development histories of other aquaculture industries, it was possible to propose a logical structure and development process for a new shrimp mariculture industry in South Carolina.

A set of assumptions related to the various components to develop a shrimp culture industry was prepared. An investigation of the available land use and geographic information permitted the establishment of a maximum land resource base on which the industry could develop.

Based on the assumption that shrimp farms would be constructed over a 10-year period, other competing land uses for this resource would reduce the area available for shrimp farms. Four levels of development were projected with a maximum of 50 % (28,116 ha) of the land resource becoming shrimp farms.

Four economic sectors—shrimp farming, grain farming, feed milling, and shrimp processing—were examined in detail. Other supporting sectors of the industry were considered as one unit. The impact of shrimp farming on the state's economy was examined from the point of minimum leakage, an ideal assumption.

From this analysis, it was apparent that the potential economic rewards justified accelerating the development process by providing increased financial assistance to the private sector.

It is interesting to note that the major international aquaculture development efforts are supported with strong backing from the World Bank, Asian Development Bank, other aid banks, and the government banks of participating countries. In the United States, our government, in general, has taken a "hands-off" position regarding direct lending assistance. Instead, development has been left to commercial banking, a sector that is less than enthusiastic about our industry. The lack of funds is a real bottleneck to accelerating industry development and restricts USA growers in competing with imported aquacultured products.

Because the potential for federal and state tax revenues is so substantial, it is recommended that state governments establish a revolving loan fund to complement the existing Farmers Home Administration loan guarantee program.

To meet the capital demands of shrimp farm development, each revolving fund should represent between 10 and 20 % of the annual capital investment required for construction and start-up operations. The balance of funds would hopefully be provided by the farmer and commercial banking. Assuming that shrimp farm land in South Carolina were developed at the rate of 2,812 ha annually, then the state's revolving loan fund should be established at a minimum level of $8 million annually (including annual loan repayments).

Although our analysis for the 1986 situation was encouraging for industry development, further advances in production yields have improved the outlook. By 1988, the Waddell Center in SC had attained experimental yields exceeding 20,000 kg live-weight/ha/crop. If these methods are adopted by the commercial sector, as we have reason to believe they will, then the economic benefit to South Carolina could conceivably increase our monetary projections by more than five fold.

In this study, many general assumptions had to be made. This process can be refined by state governments and economists for their respective areas, using our general model. Our goal is to accelerate the development process and help to establish a competitive industry while coastal property is still available in the United States.

LITERATURE CITED

Catfish Farmers of America. 1987. Farm raised catfish national facts. Catfish Farmers of America, Jackson, Mississippi.

Cooperative Extension Service. 1988. Enterprise budgets for soybeans, wheat, and corn. Cooperative Extension Service, Department of Agricultural Economics and Rural Sociology, Clemson University, Clemson, South Carolina.

Fenucci, J. L., Z. P. Zein-Eldin, and A. L. Lawrence. 1980. The nutritional response of two penaeid species to various levels of squid meal in a prepared feed. Proceedings of the World Mariculture Society 11: 405-409.

French, B., L. Cotsapas, and M.O. Hayes. 1988. Shrimp mariculture: a concept for economic development In South Carolina. Unpublished Ms. RPI International, Inc., Columbia, SC.

Goldkist Milling Company, Gaston, South Carolina. 26 January 1988. Personal communication.

Lee, K. C. January 1986. A study of the Mississippi input-output model. Mississippi Research and Development Center. Jackson, MS.

Lee, K. C. February 1988. Mississippi Research and Development Center, Jackson, Mississippi. Personal communication.

National Academy of Sciences. 1983. Nutrient requirements of warmwater fishes and shellfishes. Revised edition. National Research Council, National Academy of Sciences, National Academy Press, Washington, D.C.

New, M. B. 1976. A review of dietary studies with shrimp and prawns. Aquaculture 9:101-144.

Pugh, J. January 1988. State of South Carolina Meteorology Laboratory. Personal communication.

Rhodes, R. J., P. A. Sandifer, and J. M. Whetstone. 1987a. A preliminary financial analysis on semi-intensive penaeid shrimp farming in South Carolina. South Carolina Marine Resources Center, Technical Report 64. Charleston, SC, USA, 23 pp.

Rhodes, R., P. Sandifer, J. Hopkin, and J. Whetstone. 1987b. Financial analysis on semi-intensive penaeid shrimp farming in South Carolina. Update to South Carolina Marine Resources Center, Technical Report 64. Charleston, SC, USA.

S.C. Agricultural Statistics Service. 1987a. South Carolina crop statistics—1985-1986; state and county data. Publication Number AE445. South Carolina Agricultural Statistics Service, Columbia, SC.

S.C. Agricultural Statistics Service. 1987b. Cash receipts from farm marketings. Publication Number AE447. South Carolina Agricultural Statistics Service, Columbia, SC.

Waldrop, E. February 1988. Mississippi Agriculture and Forestry Experimental Station. Personal communication.

Welborn, T. L., Jr. 1987. Catfish farmer's handbook. Publication 1549. Extension Service of Mississippi State University.

Williams, G. 1988. Catfish Farmers of America, Jackson, MS. Personal communication.